DATA

SCIENCE

INTERVIEW

GUIDE

ACE-PREP

ABOUT THE AUTHOR

ACE-PREP

ACE PREP are researchers

Based in London, England.

The ACE-PREP is a collective; we work with the most senior academic researchers, writers and knowledge makers.

We are in the changing lives business.

DATA

SCIENCE

INTERVIEW

GUIDE

"An investment in knowledge pays the best interest"

- Benjamin Franklin.

MASTERSHIP BOOKS

UK | USA | Canada | Ireland | Australia
India | New Zealand | South Africa | China

Mastership Books is part of the United Arts Publishing House group of
companies based in London, England, UK.

First published by Mastership Books (London, UK), 2022

I S B N: 978-1-915002-10-5

Cover design by Rich © United Arts Publishing (UK)
Text and internal design by Rich © United Arts Publishing (UK)
Image credits reserved.
Colour separation by Spitting Image Design Studio
Printed and bound in Great Britain

National Publications Association of Britain
London, England, United Kingdom.
Paper design UAP

ISBN: 978-1-915002-10-5 - (paperback)

A723.5

Title: Data Science Interview Guide

Design, Bound & Printed:

London, England,
Great Britain.

CONTENTS

INTRODUCTION

Meaning of Data Science

Data science is an interdisciplinary subject that mines raw data, analyzes it, and creates patterns from which valuable insights may be extracted. Data science is based on a foundation of statistics, computer science, machine learning, deep learning, data analysis, data visualization, and a variety of other technologies.

Because of the importance of data, data science has grown in popularity in recent times. Data is seen as the new oil, which may be extremely useful to all parties when correctly examined and utilized. Not only that but a data scientist is exposed to work in many disciplines, tackling real-world challenges using cutting-edge technologies. The most popular real-time use is fast food delivery in applications like Uber Eats, which assists the delivery worker by showing the quickest path from the restaurant to the location. Data Science is also utilized in item recommendation algorithms on e-commerce sites such as Amazon, Flipkart, and others, which suggest items to users according to their search history. Data Science is also becoming extremely prevalent in detecting fraud involved in credit-based financial applications, not simply recommendation systems. While solving challenges that assist drive business and strategic goals, a skilled data scientist can understand data, innovate, and bring forth creativity. As a result, it has fast become one of the most highly paid and sought after employment roles of the twenty-first century.

Background Interview Questions and Solutions
1. What exactly does the term "Data Science" mean?

Data Science is an interdisciplinary discipline that encompasses a variety of scientific procedures, algorithms, tools, and machine learning algorithms that work together to uncover common patterns and gain useful insights from raw input data using statistical and mathematical analysis.

Gathering business needs and related data is the first step; data cleansing, data staging, data warehousing, and data architecture are all procedures in the data acquisition process. Exploring, mining, and analyzing data are all tasks that data processing does, and the results may then be utilized to provide a summary of the data's insights.

Following the exploratory phases, the cleansed data is exposed to many algorithms, such as predictive analysis, regression, text mining, pattern recognition, and so on, depending on the needs. In the final last stage, the outcomes are aesthetically appealingly when conveyed to the business. This is where the ability to see data, report on it, and use other business intelligence tools come into play.

2. What is the difference between data science and data analytics?

Data science is altering data using a variety of technical analysis approaches to derive useful insights that data analysts may apply to their business scenarios.

Data analytics is concerned with verifying current hypotheses and facts and answering questions for a more efficient and successful business decision-making process.

Data Science fosters innovation by providing answers to questions that help people make connections and solve challenges in the future. Data analytics is concerned with removing current meaning from past context, whereas data science is concerned with predictive modelling.

Data science is a wide topic that employs a variety of mathematical and scientific tools and methods to solve complicated issues. In contrast, data analytics is a more focused area that employs fewer statistical and visualization techniques to solve particular problems.

3. What are some of the strategies utilized for sampling? What is the major advantage of sampling?

Data analysis cannot be done on an entire amount of data at a time, especially when it concerns bigger datasets. It becomes important to obtain data samples that can represent the full population and then analyse it. While doing this, it is vital to properly choose sample data out of the enormous data that represents the complete dataset.

There are two types of sampling procedures depending on the engagement of statistics, they are:

- Non-Probability sampling techniques: Convenience sampling, Quota sampling, snowball sampling, etc.

- Probability sampling techniques: Simple random sampling, clustered sampling, stratified sampling.

4. List down the criteria for Overfitting and Underfitting

Overfitting: The model works well just on the sample training data. Any new data is supplied as input to the model fails to generate any result. These situations emerge owing to low bias and large variation in the model. Decision trees are usually prone to overfitting.

Underfitting: Here, the model is very simple in that it cannot find the proper connection in the data, and consequently, it does not perform well on the test data. This might arise owing to excessive bias and low variance. Underfitting is more common in linear regression.

5. Distinguish between data in long and wide formats.

Data in a long format

- Each row of the data reflects a subject's one-time information. Each subject's data would be organized in different/multiple rows.

- When viewing rows as groupings, the data may be identified.

- This data format is commonly used in R analysis and for writing to log files at the end of each experiment.

Data in a Wide Format

- The repeated replies of a subject are divided into various columns in this example.

- By viewing columns as groups, the data may be identified.

- This data format is most widely used in stats programs for repeated measures ANOVAs and is seldom utilized in R analysis.

6. What is the difference between Eigenvectors and Eigenvalues?

Eigenvectors are column vectors of unit vectors with a length/magnitude of 1; they are also known as right vectors. Eigenvalues are coefficients applied to eigenvectors to varying length or magnitude values. Eigen decomposition is the process of breaking down a matrix into Eigenvectors and Eigenvalues. These are then utilized in machine learning approaches such as PCA (Principal Component Analysis) to extract useful information from a matrix.

7. What does it mean to have high and low p-values?

A p-value measures the possibility of getting outcomes equal to or greater than those obtained under a certain hypothesis, provided the null hypothesis is true. This indicates the likelihood that the observed discrepancy happened by coincidence.

When the p-value is less than 0.05, we say have a low p-value, the null hypothesis may be rejected, and the data is unlikely to be true null. A high p-value indicates the strength in support of the null hypothesis, i.e., values greater than 0.05, indicating that the data is true null. The hypothesis can go either way with a p-value of 0.05.

8. When to do re-sampling?

Re-sampling is a data sampling procedure that improves accuracy and quantifies the uncertainty of population characteristics. It is observed that the model is efficient by training it on different patterns in a dataset to guarantee that variances are taken care of. It's also done when models need to be verified using random subsets or tests with labels substituted on data points.

9. What does it mean to have "imbalanced data"?

A data is highly imbalanced when the data is unevenly distributed across several categories. These datasets cause a performance problem in the model and inaccuracies.

10. Do the predicted value, and the mean value varies in any way?

Although there aren't many variations between these two, it's worth noting that they're employed in different situations. In general, the mean value talks about the probability distribution; in contrast, the anticipated value is used when dealing with random variables.

11. What does Survivorship bias mean to you?

Due to a lack of prominence, this bias refers to the logical fallacy of focusing on parts that survived a procedure while missing others that did not. This bias can lead to incorrect conclusions being drawn.

12. Define key performance indicators (KPIs), lift, model fitting, robustness, and design of experiment (DOE).

KPI is a metric that assesses how successfully a company meets its goals.

Lift measures the target model's performance compared to a random choice model. The lift represents how well the model predicts compared to if there was no model.

Model fitting measures how well the model under consideration matches the data.

Robustness refers to the system's capacity to successfully handle variations and variances.

DOE refers to the work of describing and explaining information variance under postulated settings to reflect factors.

13. Identify confounding variables

Another name for confounding variables is confounders. They are extraneous variables that impact both independent and dependent variables, generating erroneous associations and mathematical correlations.

14. What distinguishes time-series issues from other regression problems?

Time series data could be considered an extension of linear regression, which uses terminology such as autocorrelation and average movement to summarize previous data of y-axis variables to forecast a better future.

Time series issues' major purpose is to forecast and predict when exact forecasts could be produced, but the determinant factors are not always known.

The presence of Time in an issue might not determine that it is a time series problem. To be determined that an issue is a time series problem, there must be a relationship between target and Time.

The observations that are closer in time are anticipated to be comparable to those far away, providing seasonality accountability. Today's weather, for example, would be in comparison to tomorrow's weather but not to weather four months from now. As a result, forecasting the weather based on historical data becomes a time series challenge.

15. What if a dataset contains variables with more than 30% missing values? How would you deal with such a dataset?

We use one of the following methods, depending on the size of the dataset:

If the datasets are minimal, the missing values are replaced with the average or mean of the remaining data. This may be done in pandas by using mean = df. Mean (), where df is the panda's data frame that contains the dataset and mean () determines the data's mean. We may use df.Fillna to fill in the missing numbers with the computed mean (mean).

The rows with missing values may be deleted from bigger datasets, and the remaining data can be utilized for data prediction.

16. What is Cross-Validation, and how does it work?

Cross-validation is a statistical approach for enhancing the performance of a model. It will be designed and evaluated with rotation using different samples of the training dataset to ensure that the model performs adequately for unknown data. The training data will be divided into groups, and the model will be tested and verified against each group in turn.

The following are the most regularly used techniques:

- Leave p-out method
- K-Fold method
- Holdout method
- Leave-one-out method

17. How do you go about tackling a data analytics project?

In general, we follow the steps below:

The first stage is to understand the company's problem or need. Then, sternly examine and evaluate the facts you've been given. If any data is missing, contact the company to clarify the needs. The following stage is to clean and prepare the data, which will then be utilized for modelling. The variables are converted, and the missing values are available here.

To acquire useful insights, run your model on the data, create meaningful visualizations, and evaluate the findings. Release the model implementation and evaluate its usefulness by tracking the outcomes and performance over a set period. Validate the model using cross-validation.

18. What is the purpose of selection bias?

Selection bias occurs when no randomization is obtained while selecting a sample subset. This bias indicates that the sample used in the analysis does not reflect the whole population being studied.

19. Why is data cleansing so important?

What method do you use to clean the data?

It is critical to have correct and clean data that contains only essential information to get good insights while running an algorithm on any data. Poor or erroneous insights and projections are frequently the product of contaminated data, resulting in disastrous consequences.

For example, while starting a large marketing campaign for a product, if our data analysis instructs us to target a product that has little demand, in reality, the campaign will almost certainly fail. As a result, the company's revenue is reduced. This is when the value of having accurate and clean data becomes apparent.

Data cleaning from many sources aid data transformation and produces data scientists may work on. Clean data improves the model's performance and results in extremely accurate predictions. When a dataset is sufficiently huge, running data on it becomes difficult. If the data is large, the data cleansing stage takes a long time (about 80% of the time), and it is impossible to include it in the model's execution. As a result, cleansing data before running the model improves the model's speed and efficiency.

Data cleaning aids in the detection and correction of structural flaws in a dataset, and it also aids in the removal of duplicates and the maintenance of data consistency.

20. What feature selection strategies are available for picking the appropriate variables for creating effective prediction models?

When utilizing a dataset in data science or machine learning techniques, it's possible that not all of the variables are required or relevant for the model to be built. To eliminate duplicating models and boost the efficiency of our model, we need to use smarter feature selection approaches.

The three primary strategies for feature selection are as follows:

- Filter Approaches: These methods only take up intrinsic attributes of features assessed using univariate statistics, not cross-validated performance. They are simple and typically quicker than wrapper approaches and need fewer processing resources.
- The Chi-Square test, Fisher's Score technique, Correlation Coefficient, Variance Threshold, Mean Absolute Difference (MAD) method, Dispersion Ratios, and more filter methods are available.
- Wrapper Approaches: These methods need a way to search greedily on all potential feature subsets, access their quality, and evaluate a classifier using the feature. The selection method uses a machine-learning algorithm that must suit the provided dataset. Wrapper approaches are divided into three categories:

Forward Selection: In this method, one feature is checked, and more features are added until a good match is found.

Backward Selection: Here, all of the characteristics are evaluated, and the ones that don't fit are removed one by one to determine which works best.

Recursive Feature Elimination: The features are examined and assessed recursively to see how well they perform.

These approaches are often computationally expensive, necessitating high-end computing resources for analysis. However, these strategies frequently result in more accurate prediction models than filter methods.

Embedded Methods

By including feature interactions while retaining appropriate computing costs, embedded techniques combine the benefits of both filter and wrapper methods. These approaches are iterative because they meticulously extract characteristics contributing to most training in each model iteration. LASSO Regularization (L1) and Random Forest Importance are two examples of embedded approaches.

21. Will reclassifying categorical variables as continuous variables improve the predictive model?

Yes! A categorical variable has no particular category ordering and can be allocated to two or more categories. Ordinal variables are comparable to categorical variables because they have a defined and consistent ordering. Treating the categorical value as just a continuous variable should result in stronger prediction models if the variable is ordinal.

22. How will you handle missing values in your data analysis?

After determining which variables contain missing values, the impact of missing values may be determined. If the data analyst can detect a pattern in these missing values, there is a potential to uncover useful information. If no patterns are detected, the missing numbers can be disregarded or replaced with default parameters such as minimum, mean, maximum, or median. The default values are assigned if the missing values are for categorical variables, and missing values are assigned mean values if the data has a normal distribution. If 80 percent of the data are missing, the analyst must decide whether to use default values or remove the variables.

23. What is the ROC Curve, and how do you make one?

The ROC (Receiver Operating Characteristic) curve depicts the difference between false-positive and true-positive rates at various thresholds. The curve is used as a surrogate for a sensitivity-specificity trade-off.

Plotting values of true-positive rates (TPR or sensitivity) against false-positive rates (FPR or (1-specificity) yields the ROC curve. TPR is the percentage of positive observations correctly predicted out of all positive observations, and the FPR reflects the fraction of observations mistakenly anticipated out of all negative observations. Take medical testing as an example: the TPR shows the rate at which patients are appropriately tested positive for an illness.

24. What are the differences between the Test and Validation sets?

The test set is used to evaluate or test the trained model's performance. It assesses the model's prediction ability. The validation set is a subset of the training set used to choose parameters to avoid overfitting the model.

25. What exactly does the kernel trick mean?

Kernel functions are extended dot product functions utilized in high-dimensional feature space to compute the dot product of vectors xx and yy. A linear classifier uses the Kernel trick approach to solve a non-linear issue by changing linearly inseparable data into separable data in higher dimensions.

26. Recognize the differences between a box plot and a histogram.

Box plots and histograms are visualizations for displaying data distributions and communicating information effectively. Histograms are examples of bar charts that depict the frequency of numerical variable values and may calculate probability distributions, variations, and outliers.

Boxplots communicate various data distribution features when the form of the distribution cannot be observed, but insights may still be gained.

Compared to histograms, they are handy for comparing numerous charts simultaneously because they take up less space.

27. How will you balance/correct data that is unbalanced?

Unbalanced data can be corrected or balanced using a variety of approaches. It is possible to expand the sample size for minority groups, and the number of samples can be reduced for classes with many data points. The following are some of the methods used to balance data:

- Utilize the proper assessment metrics: It's critical to use the right evaluation metrics that give useful information while dealing with unbalanced data.
- Specificity/Precision: The number of relevant examples that have been chosen.
- Sensitivity: Indicates how many relevant cases were chosen. The F1 score represents the harmonic mean of accuracy and sensitivity, and the MCC represents the correlation coefficient between observed and anticipated binary classifications (Matthews's correlation coefficient).

The AUC (Area Under the Curve) measures the relationship between true-positive and false-positive rates.

Set of Instructions

- Resampling

Working on obtaining multiple datasets may also be used to balance data, which can be accomplished by resampling.

- Under-sampling

When the data amount is adequate, this balances the data by lowering the size of the plentiful class. A new balanced dataset may be obtained, which can be used for further modelling.

- Over-sampling

When the amount of data available is insufficient, this method is utilized. This strategy attempts to balance the dataset by increasing the sample size. Instead of getting rid of excessive samples, repetition, bootstrapping, and other approaches are used to produce and introduce fresh samples.

- Correctly do K-fold cross-validation

When employing over-sampling, cross-validation must be done correctly. Cross-validation should be performed before over-sampling since doing it afterward would be equivalent to overfitting the model to obtain a certain outcome. Data is resampled many times with varied ratios to circumvent this.

28. Random forest or many decision trees: which is better?

Because random forests are an ensemble approach that guarantees numerous weak decision trees learn forcefully, they are far more robust, accurate, and less prone to overfitting than multiple decision trees.

Careers in Data Science
Database Manager

According to Indeed, a database manager is responsible for "maintaining an organization's databases, including detecting and repairing errors, simplifying information, and reporting," according to Indeed. They also assist in selecting appropriate hardware and software solutions for their company's requirements.

Data Analyst

For businesses, data analysts collect and analyze enormous volumes of data before making suggestions based on their findings. They can improve operations, decrease costs, discover patterns, and increase efficiency in many industries, including healthcare, IT, professional sports, and finance.

Data Modeler

Systems analysts who develop computer databases that transform complex corporate data into useful computer systems are data modelers. Data modelers collaborate with data architects to create databases that fulfil organizational goals using conceptual, physical, and logical data models.

Machine Learning Engineer

A machine learning engineer is an IT professional specializing in studying, developing, and constructing self-running artificial intelligence systems to automate predictive models.

Machine learning engineers build and develop AI algorithms that can learn and make predictions, which machine learning is all about.

Business Intelligence Developer

Business intelligence developers construct systems and applications that allow users to locate and interact with their required information for an organization.

Dashboards, search functions, data modelling, and data visualization apps are examples of this. Data scientists and user experience best practices must be well-understood by BI developers.

1

MASTERING THE BASICS

Statistics

Statistics is a fascinating discipline that significantly influences today's computing world and handling extensive data. Countless businesses are pouring billions of dollars on statistics and analytics, and Statistics allows for establishing many employees in this industry and increasing competitiveness. To assist you with your Statistics interview, we've compiled a list of interview questions and answers to show you how to approach and respond to questions successfully. You'll be more prepared for the interview you're preparing for as a result of your preparation.

Basic Interview Questions on Statistics

1. What criteria do we use to determine the statistical importance of an instance?

The statistical significance of insight is determined by hypothesis testing. The null and alternate hypotheses are provided, and the p-value is computed to explain further. We considered a null hypothesis true after computing the p-value, and the values were calculated. The alpha value, which indicates importance, is changed to fine-tune the outcome. The null hypothesis is rejected if the p-value is smaller than the alpha. As a consequence, the given result is statistically significant.

2. What are the applications of long-tail distributions?

A long-tailed distribution is when the tail progressively diminishes as the curve progresses to the finish. The usage of long-tailed distributions is exemplified by the Pareto principle and the product sales distribution, and it's also famous for classification and regression difficulties.

3. What is the definition of the central limit theorem, and what is its application?

The central limit theorem asserts that when the sample size changes without changing the form of the population distribution, the normal distribution is obtained. The central limit theorem is crucial since it is commonly utilized in hypothesis testing and precisely calculating confidence intervals.

4. In statistics, what do we understand by observational and experimental data?

Data from observational studies, in which variables are examined to see a link, is referred to as observational data. Experimental data comes from investigations in which specific factors are kept constant to examine any disparity in the results.

5. What does mean imputation for missing data means? What are its disadvantages?

Mean imputation is a seldom-used technique that involves replacing null values in a dataset with the data's mean. It's a terrible approach since it removes any accountability for feature correlation. This also indicates that the data will have low variance and a higher bias, reducing the model's accuracy and narrowing confidence intervals.

6. What is the definition of an outlier, and how do we recognize one in a dataset?

Data points that differ significantly from the rest of the dataset are called outliers. Depending on the learning process, an outlier can significantly reduce a model's accuracy and efficiency.

Two strategies are used to identify outliers:

- Interquartile range (IQR)
- Standard deviation/z-score

7. In statistics, how are missing data treated?

In Statistics, there are several options for dealing with missing data:

- Missing values prediction
- Individual (one-of-a-kind) value assignment
- Rows with missing data should be deleted
- Imputation by use of a mean or median value
- Using random forests to help fill in the blanks

8. What is exploratory data analysis, and how does it differ from other types of data analysis?

Investigating data to comprehend it better is known as exploratory data analysis. Initial investigations are carried out to identify patterns, detect anomalies, test hypotheses, and confirm correct assumptions.

9. What is selection bias, and what does it imply?

The phenomenon of selection bias refers to the non-random selection of individual or grouped data to undertake analysis and better understand model functionality. If proper randomization is not performed, the sample will not correctly represent the population.

10. What are the many kinds of statistical selection bias?

As indicated below, there are different kinds of selection bias:

- Protopathic bias
- Observer selection
- Attrition
- Sampling bias
- Time intervals

11. What is the definition of an inlier?

An inlier is a data point on the same level as the rest of the dataset. As opposed to an outlier, finding an inlier in a dataset is more challenging because it requires external data. Outliers diminish model accuracy, and inliers do the same. As a result, they're also eliminated if found in the data. This is primarily done to ensure that the model is always accurate.

13. Describe a situation in which the median is superior to the mean.

When some outliers might skew data either favorably or negatively, the median is preferable since it offers an appropriate assessment in this instance.

14. Could you provide an example of a root cause analysis?

As the name implies, root cause analysis is a problem-solving technique that identifies the problem's fundamental cause. For instance, if a city's greater crime rate is directly linked to higher red-colored shirt sales, this indicates that the two variables are positively related. However, this does not imply that one is responsible for the other.

A/B testing or hypothesis testing may always be used to assess causality.

15. What does the term "six sigma" mean?

Six sigma is a quality assurance approach frequently used in statistics to enhance procedures and functionality while working with data. A process is called six sigma when 99.99966 percent of the model's outputs are defect-free.

16. What is the definition of DOE?

In statistics, DOE stands for "Design of Experiments." The task design specifies the data and varies when the independent input factors change.

17. Which of the following data types does not have a log-normal or Gaussian distribution?

There are no log-normal or Gaussian distributions in exponential distributions, and in reality, these distributions do not exist for categorical data of any kind. Typical examples are the duration of a phone call, the time until the next earthquake, and so on.

18. What does the five-number summary mean in Statistics?

As seen below, the five-number summary is a measure of five entities that encompass the complete range of data:

- Upper quartile (Q3)
- High extreme (Max)
- Median
- Low extreme (Min)
- The first quartile (Q1)

19. What is the definition of the Pareto principle?

The Pareto principle, commonly known as the 80/20 rule, states that

80% of the results come from 20% of the causes in a given experiment. The observation that 80 percent of peas originate from 20% of pea plants on a farm is a basic example of the Pareto principle.

Probability

Data scientists and machine learning engineers rely on probability theory to undertake statistical analysis of their data. Testing for probability abilities is a suitable proxy metric for organizations to assess analytical thinking and intellect since probability is also strikingly unintuitive.

Probability theory is used in different situations, including coin flips, choosing random numbers, and determining the likelihood that patients would test positive for a disease. Understanding probability might mean the difference between gaining your ideal job and having to go back to square one if you're a data scientist.

Interview Questions on Probability Concepts

These probability questions are meant to test your understanding of probability theory on a conceptual level. You might be tested on the different forms of distributions, the Central Limit Theorem or the application of Bayes' Theorem. This issue requires proper probability theory understanding and explaining this information to a layperson.

1. How do you distinguish between the Bernoulli and binomial distributions?

The Bernoulli distribution simulates one trial of an experiment with just two possible outcomes, whereas the binomial distribution simulates n trials.

2. Describe how a probability distribution might be non-normal and provide an example.

The probability distribution is not normal if most observations do not cluster around the mean, creating the bell curve. A uniform probability distribution is an example of a non-normal probability distribution, in which all values are equally likely to occur within a particular range.

4. How can you tell the difference between correlation and covariance? Give a specific example.

Covariance can be any numeric value, but correlation can only be between -1 (strong negative correlation) and 1 (strong positive correlation) (strong direct correlation). As a result, a link between two variables may appear to have a high covariance but a low correlation value.

5. How are the Central Limit Theorem and the Law of Large Numbers different?

The Law of Large Numbers states that a "sample mean" is an unbiased estimator of the population mean and that the error of that mean decreases as the sample size grows. In contrast, the Central Limit Theorem states that as a sample size n grows large, the normal distribution can approximate its distribution.

6. What is the definition of an unbiased estimator? Give a layperson an example.

An accurate statistic used to estimate a population parameter is an unbiased estimator. An example is using a sample of 1000 voters in a political poll to assess the overall voting population, and there is nothing like an utterly objective estimator.

7. Assume that the chance of finding a particular object X at location A is 0.6 and that finding it at location B is 0.8. What is the likelihood of finding item X in places A or B?

Let us begin by defining our probabilities:

P(Item at location A) = P(A) = 0.6

P(Item at location B) = P(B) = 0.8

We want to understand how likely it is that item X will be found on the internet in this city. The likelihood that item X is at location A or location B may be calculated from the question. We can describe this probability in equation form since our occurrences are not mutually exclusive: P(A or B) = P(A or B) (AUB)

8. Assume you have a deck of 500 cards with numbers ranging from 1 to 500. What is the likelihood of each following card being larger than the previously drawn card if all the cards are mixed randomly, and you are asked to choose three cards one at a time?

Consider this a sample space problem, with all other specifics ignored. We may suppose that if someone selects three distinct numbered unique cards at random without replacement, there will be low, medium, and high cards.

Let's pretend we drew the numbers 1, 2, and 3 to make things easier. In our case, the winning scenario would be if we pulled (1,2,3) in that precise order. But what is the complete spectrum of possible outcomes?

9. Assume you have one function, which gives a random number between a minimum and maximum value, N and M. Then take the output of that function and use it to calculate the total value of another random number generator with the same minimum value N. How would the resulting sample distribution be spread? What

would the second function's anticipated value be?

Let X be the first run's outcome, and Y be the second run's result. Because the integer output is "random" and no other information is provided, we may infer that any integers between N and M have an equal chance of being chosen. As a result, X and Y are discrete uniform random variables with N & M and N & X limits, respectively.

10. An equilateral triangle has three zebras seated on each corner. Each zebra chooses a direction at random and only sprints along the triangle's outline to either of the triangle's opposing edges. Is there a chance that none of the zebras will collide?

Assume that all of the zebras are arranged in an equilateral triangle. If they're sprinting down the outline to either edge, they have two alternatives for going in. Let's compute the chances that they won't collide, given that the scenario is random. In reality, there are only two options. The zebras will run in either a clockwise or counter-clockwise motion.

Let's see what the probability is for each one. The likelihood that each zebra will choose to go clockwise is the product of the number of zebras who opt to travel clockwise. Given two options (clockwise or counter-clockwise), that would be $1/2 * 1/2 * 1/2 = 1/8$.

Every zebra has the same 1/8 chance of traveling counter-clockwise. As a result, we obtain the proper probability of 1/4 or 25% if we add the probabilities together.

11. You contact three random pals in Seattle and ask them if it's raining on their own. Each of your pals has a two-thirds probability of giving you the truth and a one-third risk of deceiving you by lying. "Yes," all three of your buddies agree, it is raining. What are the chances that it's raining in Seattle right now?

According to the outcome of the Frequentist method, if you repeat the trials with your friends, there is one occurrence in which all three of your friends lied inside those 27 trials.

However, because your friends all provided the same response, you're not interested in all 27 trials, which would include occurrences when your friends' replies differed.

12. You have 576 times to flip a fair coin. Calculate the chance of flipping at least 312 heads without using a calculator.

This question needs a little memory. Given that we have to predict the number of heads out of some trials, we may deduce that it's a binomial distribution problem at first look. As a result, for each test, we'll employ a binomial distribution with n trials and a probability of success of p. The probability for success (a fair coin has a 0.5 chance of landing heads or tails) multiplied by the total number of trials is the anticipated number of heads for a binomial distribution (576). As a result, our coin flips are projected to turn up heads 288 times.

13. You are handed a neutral coin, and you are to toss the coin until it lands on either Heads Heads Tails (HHT) or Heads Tails Tails (HTT). Is it more probable that one will appear first? If so, which one and how likely is it?

Given the two circumstances, we may conclude that both sequences need H to come first. The chance of HHT is now equal to 1/2 after H occurs. What is the reason behind this? Because all you need for HHT in this circumstance is one H. Because we are flipping the coin in series until we observe a string of HHT or HTT in a row, the coin does not reset. The fact that the initial letter is H enhances the likelihood of HHT rather than HTT.

Linear Algebra

From the notations used to describe the operation of algorithms to the implementation of algorithms in code, linear algebra is a vital basis in the study of machine learning and data science. Check out popular Linear Algebra Interview Questions that any ML engineer and data scientist should know before their following data science interview.

Interview Questions on Linear Algebra

1. Under what circumstances does the inverse of a diagonal matrix exist?

If all diagonal elements are non-zero, the inversion of a square diagonal matrix exists. If this is the case, the inverse is derived by substituting the reciprocal of each diagonal element.

2. What does Ax = bAx=b stand for? When does Ax = b have a unique solution?

Ax = b is a set of linear equations written in matrix form, in which:

A is the order m x n coefficient matrix, and x is the order n x 1 incognite variables vector.

The constants create the vector b, which has the order m x 1.

If and only if the system Ax = b has a unique solution.

n rank[A|b] = n rank[A|b] = n

n rank[A|b] = n rank[A|b] = n

The matrix A|b is matrix A with b attached as an additional column, hence rank[A]=rank[Ab]=n.

3. What is the process for diagonalizing a matrix?

To obtain the diagonal matrix D of an nxn matrix A, we must do the following:

- Determine A's characteristic polynomial.
- To get the eigenvalues of A, find the roots of the characteristic polynomial.
- Find the corresponding eigenvectors for each of A's eigenvalues.

The matrix is not diagonalizable if the total number of eigenvectors m determined in step 3 does not equal n (the number of rows and columns in A), but if m = n, the diagonal matrix D is provided by: bf D = P-1 A P, D=P 1 AP, where P is defined as a matrix whose columns are the eigenvectors of the matrix A.

4. Find the definition of positive definite, negative definite, positive semi-definite, and negative semi-definite matrices?

A positive definite matrix (PsDM) is a symmetric matrix M in which the number ztMz is positive for each non-zero column vector z.

A symmetric matrix M is a positive semi-definite matrix if the number ztMz is positive or zero for every non-zero column vector z. Negative semi-definite matrices and negative definite matrices are defined in the same way.

Because each matrix may be linked to the quadratic equation ztMz, these matrices aid in solving optimization issues, a positive definite matrix M, for example, implies a convex function, ensuring the existence of the global minimum. This allows us to solve the optimization issue with the Hessian matrix, and negative definite matrices are subject to the same considerations.

5. How does Linear Algebra relate to broadcasting?

Broadcasting is a technique for easing element-by-element operations based on dimension constraints. If the relevant dimensions in each matrix (rows versus rows, columns versus columns) match the following conditions, we say two matrices are compatible for broadcasting.

The measurements are the same, or one of the dimensions is the same size. Broadcasting works by duplicating the smaller array to make it the same size and dimensions as the bigger array. This approach was initially created for NumPy, but it has now been adopted by some other numerical computing libraries, including Theano, TensorFlow, and Octave.

6. What is an Orthogonal Matrix?

An orthogonal matrix is a square matrix with columns and rows of orthonormal unit vectors, such as perpendicular and length or magnitude. It's formally defined as follows:

$$Q^t Q = Q Q^t = I$$

$$Qt, Q = QQ, \quad t = I$$

Where Q stands for the orthogonal matrix, Qt for the transpose of Q, and I for the identity matrix. We can observe from the definition above that

$$Q^{-1} = Q^t$$

$$Q-1 = Qt$$

As a result, the orthogonal matrix is favored since its inverse is computed as merely its transpose, computationally inexpensive, and stable. Then you'll need to recall that the binomial distribution's standard deviation is sqrt(n*p*(1-p)).

2

PYTHON

Guido van Rossum created Python, initially released on February 20, 1991. It is one of the most popular and well-liked programming languages, and since it is interpreted, it allows for the incorporation of dynamic semantics. It's also a free and open-source language with straightforward syntax, making it simple for programmers to learn Python. Python also allows object-oriented programming and is the most widely used programming language.

Python's popularity is skyrocketing, thanks to its ease of use and ability to perform several functions with fewer lines of code. Because of its capacity to handle strong calculations utilizing powerful libraries, Python is also utilized in Machine Learning, Artificial Intelligence, Web Development, Web Scraping, and many other disciplines. As a result, Python developers are in high demand in India and worldwide, and companies are eager to provide these professionals with incredible advantages and privileges.

Interview Questions in Python
1. What exactly is Python?

Python is a general-purpose, high-level, interpreted programming language. The correct tools/libraries may be used to construct practically any application because it is a general-purpose language. Python also has features like objects, modules, threads, exception handling, and automated memory management, which aid in modelling real-world issues and developing programs to solve them.

2. What are the advantages of Python?

Python is a general-purpose programming language with a simple, easy-to-learn syntax that prioritizes readability and lowers program maintenance costs. Furthermore, the language is scriptable, open-source, and enables third-party packages, promoting modularity and code reuse.

Its high-level data structures, along with the dynamic type and dynamic binding, have attracted a large developer community for Rapid Application Development and deployment.

3. What is the definition of dynamically typed language?

We must first learn about typing before comprehending a dynamically typed language. In computer languages, typing refers to type-checking. Because these languages don't allow for "type-coercion," "1" + 2 will result in a type error in a strongly-typed language like Python (implicit conversion of data types). On the other hand, a weakly-typed language, such as JavaScript, will simply return "12" as a result.

There are two steps to type-checking

- Static - Data Types are checked before execution.
- Dynamic - Data Types are checked during execution.

Python is an interpreted language that executes each statement line by line. Thus type-checking happens in real-time while the program is running, and python is a Dynamically Typed Language as a result.

4. What is the definition of an Interpreted Language?

The sentences in an Interpreted language are executed line by line. Interpreted languages include Python, JavaScript, R, PHP, and Ruby, to name just a few. An interpreted language program executes straight from the source code without a compilation phase.

5. What is the meaning of PEP 8, and how significant is it?

Python Enhancement Proposal (PEP) is an acronym for Python Enhancement Proposal. A Python Extension Protocol (PEP) is an official design document that provides information to the Python community or describes a new feature or procedure for Python. PEP 8 is particularly important since it outlines the Python code style rules. Contributing to the Python open-source community appears to need a serious and tight adherence to these stylistic rules.

6. What is the definition scope in Python?

In Python, each object has its scope. In Python, a scope is a block of code in which an object is still relevant. Namespaces uniquely identify all the objects in a program. On the other hand, these namespaces have a scope set for them, allowing you to utilize their objects without any prefix. The following are a few instances of scope produced during Python code execution:

- Those local objects available in a particular function are a local scope.
- A global scope refers to the items that have been available from the beginning of the code execution.
- The global objects of the current module that are available in the

program are referred to as a module-level scope.

- An outermost scope refers to all of the program's built-in names. The items in this scope are searched last to discover the name reference.

Note: Keywords like global can sync local scope items with global scope ones.

9. What is the meaning of pass in Python?

In Python, the pass keyword denotes a null operation. It is commonly used to fill in blank blocks of code that may execute during runtime but has not yet been written. We may encounter issues during code execution if we don't use the pass statement in the following code.

20. How does Python handle memory?

The Python Memory Manager is in charge of memory management in Python. The memory allotted by the manager is in the form of a Python-only private heap area. This heap holds all Python objects, and because it is private, it is unavailable to the programmer. Python does, however, have several basic API methods for working with the private memory area. Python also features a built-in garbage collection system that recycles unneeded memory for the private heap area.

21. What are namespaces in Python? What is their purpose?

In Python, a namespace ensures that object names are unique and used without conflict. These namespaces are implemented in Python as dictionaries with a 'name as key' and a corresponding 'object as value.' Due to this, multiple namespaces can use the same name and map it to a different object. Here are a few instances of namespaces:

- Within a function, local names are stored in the Local Namespace. A temporary namespace is formed when a function is called, removed when the function returns.

- The names of various imported packages/modules used in the current project are stored in the Global Namespace. When the package is imported into the script, this namespace is generated and persists until executed.
- Built-in Namespace contains essential Python built-in functions and built-in names for different sorts of exceptions.
- The lifespan of a namespace is determined by the scope of objects to which it is assigned. The lifespan of a namespace comes to an end when the scope of an object expires. As a result, accessing inner namespace objects from an outside namespace is not feasible.

22. What is Python's Scope Resolution?

Objects with the same name but distinct functions exist inside the same scope. In certain instances, Python's scope resolution kicks in immediately. Here are a few examples of similar behavior:

Many functions in the Python modules 'math' and 'cmath' are shared by both - log10(), acos(), exp(), and so on. It is important to prefix them with their corresponding module, such as math.exp() and cmath.exp(), to overcome this problem.

Consider the code below, where an object temp is set to 10 globally and subsequently to 20 when the function is called. The function call, however, did not affect the global temperature value. Python draws a clear distinction between global and local variables, interpreting their namespaces as distinct identities.

23. Explain the definition of decorators in Python?

Decorators in Python are simply functions that add functionality to an existing Python function without affecting the function's structure. In Python, they are represented by the name @decorator name and are invoked from the bottom up.

The elegance of decorators comes in the fact that, in addition to adding functionality to the method's output, they may also accept parameters for functions and change them before delivering them to the function. The inner nested function, i.e., the 'wrapper' function, is crucial in this case, and it's in place to enforce encapsulation and, as a result, keep itself out of the global scope.

24. What are the definitions of dict and list comprehensions?

Python comprehensions, like decorators, are syntactic sugar structures that aid in the construction of changed and filtered lists, dictionaries, and sets from a given list, dictionary, or set. Using comprehensions saves a lot of effort and allows you to write less verbose code (containing more lines of code). Consider the following scenarios in which comprehensions might be highly beneficial:

- Performing math operations throughout the full list
- Using conditional filtering to filter the entire list
- Multiple lists can be combined into one using comprehensions, which allow for many iterators and hence can be used to combine multiple lists into one.
- Taking a multi-dimensional list and flattening it
- A similar strategy of nested iterators (as seen before) can be used to flatten a multi-dimensional list or operate on its inner members.

25. What is the definition of lambda in Python? What is the purpose of it?

In Python, a lambda function is an anonymous function that can take any number of parameters but only have one expression. It's typically utilized when an anonymous function is required for a brief time.

Lambda functions can be applied in two different ways:

To assign lambda functions to a variable, do the following:

```
mul = lambda a, b : a * b
print(mul(2, 5)) # output => 10
Wrapping lambda functions inside another function:
def. myWrapper(n):
 return lambda a : a * n
mulFive = myWrapper(5)
print(mulFive(2)) # output => 10
```

26. In Python, how do you make a copy of an object?

The assignment statement (= operator) in Python doesn't duplicate objects. Instead, it establishes a connection between the existing object and the name of the target variable. In Python, we must use the copy module to make copies of an object. Furthermore, the copy module provides two options for producing copies of a given object –

A bit-wise copy of an object is called a shallow copy. The values of the cloned object are an identical replica of the original object's values. If one of the variables references another object, just the references to that object are copied. Deep Copy recursively replicates all values from source to destination object, including the objects referenced by the source object.

28. What are the definitions of pickling and unpickling?

"Serialization out of the box" is a feature that comes standard with the Python library. Serializing an object means converting it into a format that can be saved to be de-serialized later to return to its original state. The pickle module is used in this case.

Pickling

In Python, the serialization process is known as pickling. In Python, any object may be serialized as a byte stream and saved as a memory file. Pickling is a compact process, but pickle items may be further compacted. Pickle also retains track of the serialized objects, which is cross-version portable. Pickle.dump is the function used in operation mentioned above ().

Unpickling

Pickling is the polar opposite of unpickling. After deserializing the byte stream, it loads the object into memory to reconstruct the objects saved in the file. Pickle.load is the function used in operation mentioned above ().

30. What is PYTHONPATH?

PYTHONPATH is an environment variable that allows you to specify extra directories in which Python will look for modules and packages. This is especially important if you want to keep Python libraries that aren't installed in the global default location.

31. What are the functions help() and dir() used for?

Python's help() method displays modules, classes, functions, keywords, and other objects. If the help() method is used without an argument, an interactive help utility is opened on the console.

The dir() function attempts to return a correct list of the object's attributes and methods. It reacts differently to various things because it seeks to produce the most relevant data rather than all of the information.

It produces a list of all characteristics included in that module for Modules/Library objects. It returns a list of all acceptable attributes and basic attributes for Class Objects. It produces a list of attributes in the current scope if no arguments are supplied.

32. How can you tell the difference between.py and.pyc files?

The source code of a program is stored in.py files. Meanwhile, the bytecode of your program is stored in the .pyc file. After compiling the.py file, we obtain bytecode (source code). For some of the files you run, .pyc files are not produced. It's solely there to hold the files you've imported—the python interpreter checks for compiled files before executing a python program. The virtual computer runs the file if it is present, and it looks for a.py file if it isn't found. It is compiled into a.pyc file and then executed by the Python Virtual Machine if it is discovered. Having a.pyc file saves you time while compiling.

33. What does the computer interpret in Python?

Python is not an interpreted or compiled language. The implementation's attribute is whether it is interpreted or compiled. Python is a bytecode (a collection of interpreter-readable instructions) that may be interpreted differently. The source code is saved with the extension .py. Python generates a set of instructions for a virtual machine from the source code. The Python interpreter is a virtual machine implementation. "Bytecode" is the name for this intermediate format. The.py source code is initially compiled into bytecode (.pyc). This bytecode can then be interpreted by the standard CPython interpreter or PyPy's JIT (Just in Time compiler).

34. In Python, how are arguments delivered by value or reference?

Pass by value: The real object is copied and passed. Changing the value of the object's duplicate does not affect the original object's value.

Pass via reference: The real object is supplied as a reference. The value of the old object will change if the value of the new object is changed.

Arguments are supplied by reference in Python, which means that a reference to the real object is passed.

3

PANDA

Pandas are the most widely used library for working with tabular data. Consider Python's version of a spreadsheet or SQL table. Structured data can be manipulated in the same way that Excel or Google Sheets can. Many machine learning and related libraries include SciPy, Scikit-learn, Statsmodels, NetworkX, and visualization libraries, including Matplotlib, Seaborn, Plotly, and others, are compatible with Pandas Data Structures. Many specialized libraries have been constructed on top of the Pandas Library, including geo-pandas, quandl, Bokeh, and others. Pandas are used extensively in many proprietary libraries for algorithmic trading, data analysis, ETL procedures, etc.

Interview Questions for Python Pandas

1. What exactly are Pandas/Python Pandas?

Pandas are a Python open-source toolkit that allows for high-performance data manipulation. Pandas get its name from "panel data," which refers to econometrics based on multidimensional data. It was created by Wes McKinney in 2008 and may be used for data analysis in Python. It can conduct the five major processes necessary for data processing and analysis, regardless of the data's origin, namely load, manipulate, prepare, model, and analyze.

2. What are the different sorts of Pandas Data Structures?

Pandas provide two data structures, Series and DataFrames, which the panda's library supports. Both of these data structures are based on the NumPy framework. A series is a one-dimensional data structure in pandas, whereas a DataFrame is two-dimensional.

3. How do you define a series in Pandas?

A Series is a one-dimensional array capable of holding many data types. The index refers to the row labels of a series. We can quickly turn a list, tuple, or dictionary into a series by utilizing the 'series' function. Multiple columns are not allowed in a Series.

4. How can the standard deviation of the Series be calculated?

The Pandas std() method is used to calculate the standard deviation of a collection of values, a DataFrame, a column, and a row.

Series.std(skipna=None, axis=None, ddof=1, level=None, numeric_only=None, **kwargs)

5. How do you define a DataFrame in Pandas?

A DataFrame is a pandas data structure that uses a two-dimensional array with labeled axes (rows and columns). A DataFrame is a typical way to store data with two indices, namely a row index, and a column index. It has the following characteristics:

Columns of heterogeneous kinds, such as int and bool, can be used, and it may be a dictionary of Series structures with indexed rows and columns. When it comes to columns, it's "columns," and when it comes to rows, it's "index."

6. What distinguishes the Pandas Library from other libraries?

The following are the essential aspects of the panda's library:

- Alignment of Data
- Efficient Memory
- Time Series
- Reshaping
- Join and merge

7. What is the purpose of reindexing in Pandas?

DataFrame is reindexed to adhere to a new index with optional filling logic. It inserts NA/NaN in areas where the values are missing from the preceding index. Unless the new index is provided as identical to the current one, the value of the copy becomes False. It returns a new object, and it is used to modify the DataFrame's rows and columns index.

10. Can you explain how to use categorical data in Pandas?

Categorical data is a Pandas data type that correlates to a categorical statistical variable. A categorical variable has a restricted number of potential values, usually fixed. Gender, place of origin, blood type, socioeconomic status, observation time, and Likert scale ratings are just a few examples. Categorical data values are either in categories or np.nan.

This data typically comes in handy in the following scenarios:

- It's handy for a string variable with a limited number of possible values. We can change a string variable to a categorical variable to save some memory.

- It is useful when a variable's lexical order differs from its logical order (one?? two?? three?). Sorting and min/max are responsible for using the logical order instead of the lexical order by converting to a categorical and specifying an order on the categories.
- Because this column should be handled as a categorical variable, it serves as a signal to other Python libraries.

12. In Pandas, how can we make a replica of the series?

The following syntax can be used to make a replica of a series:

- Series.copy(deep=True)
- pandas.Series.copy

The statements above create a deep copy, which contains a copy of the data and the indices. If we set deep to False, neither the indices nor the data will be copied.

13. How can I rename a Pandas DataFrame's index or columns?

You may use the .rename method to change the values of DataFrame's columns or index values.

14. What is the correct way to iterate over a Pandas DataFrame?

By combining a loop with an iterrows() function on the DataFrame, you may iterate over the rows of the DataFrame.

15. How Do I Remove Indices, Rows, and Columns from a Pandas Data Frame?

You must perform the following if you wish to delete the index from the DataFrame:

Dataframe's Index Reset

- To delete the index name, run del df.index.name.
- Reset the index and drop the duplicate values from the index column to remove duplicate index values.
- With a row, you may remove an index.

Getting Rid of a Column in your Dataframe

- The drop() function may remove a column from a DataFrame.
- The axis option given to the drop() function is either 0 to indicate the rows or 1 to indicate the columns to be dropped.
- To remove the column without reassigning the DataFrame, pass the argument in place and set it to True.
- The drop duplicates() function may also remove duplicate values from a column.

Getting Rid of a Row in your Dataframe

- We may delete duplicate rows from the DataFrame by calling df.drop duplicates().
- The drop() function may indicate the index of the rows to be removed from the DataFrame.

16. What is a NumPy array in Pandas?

Numerical Python (Numpy) is a Python module that allows you to do different numerical computations and handle multidimensional and single-dimensional array items. Numpy arrays are quicker than regular Python arrays for computations.

17. What is the best way to transform a DataFrame into a NumPy array?

We can convert Pandas DataFrame to NumPy arrays to conduct various high-level mathematical procedures. The DataFrame.to NumPy() method is used.

The DataFrame.to_numpy() function is used to the DataFrame which returns the numpy ndarray. DataFrame.back to_the numpy(dtype=None, copy=False).

18. What is the best way to convert a DataFrame into an Excel file?

Using the to excel() method, we can export the DataFrame to an excel file. We must mention the destination filename to write a single object to an excel file. If we wish to write too many sheets, we must build an ExcelWriter object with the destination filename and the sheet in the file that we want to write to.

19. What is the meaning of Time Series in panda?

Time series data is regarded as an important source of information for developing a strategy that many organizations may use. It contains a lot of facts about the time, from the traditional banking business to the education industry. Time series forecasting is a machine learning model that deals with Time Series data to predict future values.

20. What is the meaning of Time Offset?

The offset defines a range of dates that meet the DateOffset's requirements. We can use Date Offsets to advance dates forward to make them legitimate.

21. How do you define Time periods?

The Time Periods reflect the length of time, such as days, years, quarters, and months. It's a class that lets us convert frequencies to periods.

Numpy Interview Questions
1. What exactly is Numpy?

NumPy is a Python-based array processing program. It includes a high-performance multidimensional array object and utilities for manipulating them. It is the most important Python module for scientific computing. An N-dimensional array object with a lot of power and sophisticated broadcasting functions.

2. What is the purpose of NumPy in Python?

NumPy is a Python module that is used for Scientific Computing. The NumPy package is used to carry out many tasks. A multidimensional array called ndarray (NumPy Array) holds the same data type values. These arrays are indexed in the same way as Sequences are, starting at zero.

3. What does Python's NumPy stand for?

NumPy (pronounced /nmpa/ (NUM-py) or /nmpi/ (NUM-pee)) is a Python library that adds support for huge, multi-dimensional arrays and matrices, as well as a vast number of high-level mathematical functions to work on these arrays.

4. Where does NumPy come into play?

NumPy is a free, open-source Python library for numerical computations. A multi-dimensional array and matrix data structures are included in NumPy, and it may execute many operations on arrays, including trigonometric, statistical, and algebraic algorithms. NumPy is a Numeric and Numarray extension.

5. Installation of Numpy into Windows?

Step 1:

Install Python on your Windows 10/8/7 computer. To begin, go to the official Python download website and download the Python executable binaries for your Windows machine.

Step 2:

Install Python using the Python executable installer.

Step 3:

Download and install pip for Windows 10/8/7.

Step 4:

Install Numpy in Python on Windows 10/8/7 using pip.

The Numpy Installation Process.

Step 1:

Open the terminal

Step 2:

Type pip install NumPy

6. What is the best way to import NumPy into Python?

Import NumPy as np

7. How can I make a one-dimensional(1D)array?

```
Num=[1,2,3]
Num = np.array(num)
Print("1d array : ",num)
```

8. How can I make a two-dimensional (2D)array?

```
Num2=[[1,2,3],[4,5,6]]
Num2 = np.array(num2)
Print("\n2d array : ",num2)
```

9. How do I make a 3D or ND array?

Num3=[[[1,2,3],[4,5,6],[7,8,9]]]
Num3 = np.array(num3)
Print("\n3d array : ",num3)

10. What is the best way to use a shape in a 1D array?

If num=[1,2,3], print('nshape of 1d',num.shape) if not defined.

11. What is the best way to use shape in a 2D array?

If not added, num2=[[1,2,3],[4,5,6]] print('nshape of 2d',num2.shape)

12. What is the best way to use shape in 3D or Nd Array?

Num3=[[[1,2,3],[4,5,6],[7,8,9]]] if not added

Print('\nshpae of 3d ',num3.shape)

13. What is the best way to identify the data type of a NumPy array?

Print('\n data type num 1 ',num.dtype)
Print('\n data type num 2 ',num2.dtype)
Print('\n data type num 3 ',num3.dtype)

14. Can you print 5 zeros?

Arr = np.zeros(5)
Print('single arrya',arr)

15. Print zeros in a two-row, three-column format?

Arr2 = np.zeros((2,3))
Print('\nprint 2 rows and 3 cols : ',arr2)

16. Is it possible to utilize eye() diagonal values?

Arr3 = np.eye(4)

Print('\ndiaglonal values : ',arr3)

17. Is it possible to utilize diag() to create a square matrix?

```
Arr3 = np.diag([1,2,3,4])
Print('\n square matrix',arr3)
```

18. Printing range Show 4 integers random numbers between 1 and 15

```
Rand_arr = np.random.randint(1,15,4)
Print('\n random number from 1 to 15 ',rand_arr)
```

19. Print a range of 1 to 100 and show four integers at random.

```
Rand_arr3 = np.random.randint(1,100,20)
Print('\n random number from 1 to 100 ',rand_arr3)
```

20. Print range between random numbers 2 rows and three columns, select integer's random numbers.

```
Rand_arr2 = np.random.randint([2,3])
Print('\n random number 2 row and 3 cols ',rand_arr2)
```

21. What is an example of the seed() function? What is the best way to utilize it? What is the purpose of seed()?

```
np.random.seed(123)
Rand_arr4 = np.random.randint(1,100,20)
Print('\nseed() showing same number only : ',rand_arr4)
```

22. What is one-dimensional indexing?

Num = np.array([5,15,25,35]) is one example.

Num = np.array([5,15,25,35])

Print('my array : ',num)

23. Print the first, last, second, and third positions.

Num = np.array([5,15,25,35]) if not added
Print('\n first position : ',num[0]) #5
Print('\n third position : ',num[2]) #25

24. How do you find the final integer in a NumPy array?

Num = np.array([5,15,25,35]) if not added
Print('\n forth position : ',num[3])

25. How can we prove it pragmatically if we don't know the last position?

Num = np.array([5,15,25,35]) if not added
Print('\n last indexing done by -1 position : ',num[-1])

4

MACHINE LEARNING

The process of teaching computer software to develop a statistical model based on data is referred to as machine learning. The purpose of machine learning (ML) is to transform data and extract essential patterns or insights from it.

Machine Learning Interview Questions

1. What was the purpose of Machine Learning?

The most straightforward response is to make our lives simpler. Many systems employed hardcoded rules of "if" and "else" choices to analyze data or change user input in the early days of "intelligent" applications. Consider a spam filter responsible for moving relevant incoming email messages to a spam folder. However, we provide enough data to learn and find patterns using machine learning algorithms.

Unlike traditional challenges, we don't need to define new rules for each machine learning problem; instead, we need to utilize the same approach but with a different dataset.

For example, if we have a history dataset of real sales statistics, we may train machine learning models to forecast future sales.

Principal Component Analysis, or PCA, is a dimensionality-reduction approach for reducing the dimensionality of big data sets by converting a large collection of variables into a smaller one that retains the majority of the information in the large set.

2. Define Supervised Learning?

Supervised learning is a machine learning technique that uses labeled training data to infer a function. A series of training examples make up the training data.

Example:

Knowing a person's height and weight might help determine their gender. The most common supervised learning algorithms are shown below.

- Support Vector Machines
- K-nearest Neighbor Algorithm and Neural Networks.
- Naive Bayes
- Regression
- Decision Trees

3. Explain Unsupervised Learning?

Unsupervised learning is a machine learning method that searches for patterns in a data set. There is no dependent variable or label to forecast in this case. Algorithms for Unsupervised Learning:

- Clustering
- Latent Variable Models and Neural Networks
- Anomaly Detection

Example:

A T-shirt clustering, for example, will be divided into "collar style and V neck style," "crew neck style," and "sleeve kinds."

4. What should you do if you're Overfitting or Underfitting?

- Overfitting occurs when a model is too well suited to training data; in this scenario, we must resample the data and evaluate model accuracy using approaches such as k-fold cross-validation.
- Whereas in Underfitting, we cannot interpret or capture patterns from the data, we must either adjust the algorithms or input more data points to the model.

5. Define Neural Network?

It's a simplified representation of the human mind. It has neurons that activate when it encounters anything comparable to the brain. The many neurons are linked by connections that allow information to travel from one neuron to the next.

6. What is the meaning of Loss Function and Cost Function? What is the main distinction between them?

When computing loss, we just consider one data point, referred to as a loss function. The cost function determines the total error for numerous data, and there isn't much difference. A loss function captures the difference between the actual and projected values for a single record, whereas a cost function aggregates the difference across the training dataset. Mean-squared error and Hinge loss are the most widely utilized loss functions. The Mean-Squared Error (MSE) measures how well our model predicted values compared to the actual values.

$MSE = \sqrt{(predicted\ value - actual\ value)2}$

Hinge loss: It is used to train the machine learning classifier, which is

$L(y) = \max(0, 1 - yy)$

Where y = -1 or 1 denotes two classes and y denotes the classifier's output form. In the equation $y = mx + b$, the most common cost function depicts the entire cost as the sum of the fixed and variable costs.

7. Define Ensemble Learning?

Ensemble learning is a strategy for creating more powerful machine learning models by combining numerous models.

There are several causes for a model's uniqueness. The following are a few reasons:

- Various Populations
- Various Hypotheses
- Various modelling approaches

We will encounter an error when working with the model's training and testing data. Bias, variation, and irreducible error are all possible causes of this inaccuracy. The model should now always exhibit a bias-variance trade-off, which we term a bias-variance trade-off. This trade-off can be accomplished by ensemble learning.

There are a variety of ensemble approaches available. However, there are two main strategies for aggregating several models:

- Bagging is a natural approach for generating new training sets from an existing one.
- Boosting is a more elegant strategy to optimize the optimum weighting scheme for a training set.

8. How do you know the Machine Learning Algorithm you should use?

It is entirely dependent on the data we have. SVM is used when the data is discrete, and we utilize linear regression if the dataset is continuous. As a result, there is no one-size-fits-all method for determining which machine learning algorithm to utilize; it all relies on exploratory data analysis (EDA). EDA is similar to "interviewing" a dataset. We do the following as part of our interview:

- Sort our variables into categories like continuous, categorical, and so on.
- Use descriptive statistics to summarize our variables.
- Use charts to visualize our variables.
- Choose one best-fit method for a dataset based on the given observations.

9. How should Outlier Values be Handled?

An outlier is a dataset observation significantly different from the rest of the dataset. The following are some of the tools that are used to find outliers.

- Z-score
- Box plot
- Scatter plot, etc.

To deal with outliers, we usually need to use one of three easy strategies:

- We can get rid of them.
- They can be labeled as outliers and added to the feature set.
- Similarly, we may change the characteristic to lessen the impact of the outlier.

10. Define Random Forest? What is the mechanism behind it?

Random forest is a machine learning approach that may be used for regression and classification. Random forest operates by merging many different tree models, and random forest creates a tree using a random sampling of the test data columns.

The procedures for creating trees in a random forest are as follows:

- Using the training data, calculate the sample size.
- Begin by creating a single node.
- From the start node, run the following algorithm:
- Stop if the number of observations is fewer than the node size.
- Choose variables at random.
- Determine which variable does the "best" job of separating the data.
- Divide the observations into two nodes.
- Run step 'a' on each of these nodes.

10. What are SVM's different Kernels?

In SVM, there are six different types of kernels, below are four of them:

- Linear kernel - When data is linearly separable.
- Polynomial kernel - When you have discrete data with no natural idea of smoothness.
- Radial basis kernel - Create a decision boundary that can separate two classes considerably better than a linear kernel.
- Sigmoid kernel - The sigmoid kernel is a neural network activation function.

11. What is Machine Learning Bias?

Data bias indicates that there is a discrepancy in the data. Inconsistency can develop for different causes, none of which are mutually exclusive.

For example, to speed up the recruiting process, a digital giant like Amazon built a single-engine that will take 100 resumes and spit out the best five candidates to employ. The program was adjusted to remove the prejudice once the business noticed it wasn't providing gender-neutral results.

11. What is the difference between regression and classification?

Classification is used to provide distinct outcomes, as well as to categorize data into specified categories. An example is classifying emails into spam and non-spam groups. Regression, on the other hand, works with continuous data. An example is Predicting stock prices at a specific period in time.

The term "classification" refers to the process of categorizing the output into a set of categories. For example, is it going to be cold or hot tomorrow? On the other hand, regression is used to forecast the connection that data reflects. An example is, what will the temperature be tomorrow?

12. What is Clustering, and how does it work?

Clustering is the process of dividing a collection of things into several groups. Objects in the same cluster should be similar to one another but not those in different clusters.

The following are some examples of clustering:

- K means clustering
- Hierarchical clustering
- Fuzzy clustering
- Density-based clustering, etc.

13. What is the best way to choose K for K-means Clustering?

Direct procedures and statistical testing methods are the two types of approaches available:

Direct Methods: It has elbows and a silhouette.

Methods of statistical testing: There are data on the gaps.

When selecting the ideal value of k, the silhouette is the most commonly utilized.

14. Define Recommender Systems

A recommendation engine is a program that predicts a user's preferences and suggests things that are likely to be of interest to them. Data for recommender systems comes from explicit user evaluations after seeing a movie or listening to music, implicit search engine inquiries and purchase histories, and other information about the users/items themselves.

15. How do you determine if a dataset is normal?

Plots can be used as a visual aid. The following are a few examples of normalcy checks:

- Shapiro-Wilk Test
- Anderson-Darling Test
- Martinez-Iglewicz Test
- Kolmogorov-Smirnov Test
- D'Agostino Skewness Test

16. Is it possible to utilize logistic regression for more than two classes?

By default, logistic regression is a binary classifier, which means it can't be used for more than two classes. It can, however, be used to solve multi-class classification issues (multinomial logistic regression)

17. Explain covariance and correlation?

Correlation is a statistical technique for determining and quantifying the quantitative relationship between two variables. The strength of a relationship between two variables is measured by correlation. Income and spending, demand and supply, and so on are examples.

Covariance is a straightforward method of determining the degree of connection between two variables. The issue with covariance is that it's difficult to compare them without normalization.

18. What is the meaning of P-value?

P-values are utilized to make a hypothesis test choice. The P-value is the least significant level where the null hypothesis may be rejected. The lower the p-value, the more probable the null hypothesis is rejected.

19. Define Parametric and Non-Parametric Models

Parametric models contain a small number of parameters. Thus all you need to know to forecast new data is the model's parameter.

Non-parametric models have no restrictions on the number of parameters they may take, giving them additional flexibility and the ability to forecast new data. You must be aware of the current status of the data and the model parameters.

20. Define Reinforcement Learning

Reinforcement learning differs from other forms of learning, such as supervised and unsupervised learning. We are not provided data or labels in reinforcement learning.

21. What is the difference between the Sigmoid and Softmax functions?

The sigmoid function is utilized for binary classification, and the sum of the probability must be 1. On the other hand, the Softmax function is utilized for multi-classification, and the total probability will be 1.

PCA Interview Questions
1. What is the Dimensionality Curse?

All of the issues that develop when working with data in more dimensions are called the curse of dimensionality. As the number of features grows, so does the number of samples, making the model increasingly complicated. Overfitting is increasingly likely as the number of characteristics increases. A machine learning model trained on a high number of features becomes overfitted as it becomes increasingly reliant on the data it was trained on, resulting in poor performance on actual data and defeating the objective. Our model will make fewer assumptions and be simpler if our training data contains fewer characteristics.

2. Why do we need to reduce dimensionality? What are the disadvantages?

The amount of features is referred to as a dimension in Machine Learning. The process of lowering the dimension of your feature collection is known as dimensionality reduction.

Dimensionality Reduction Benefits

- With less misleading data, model accuracy increases.
- Less computation is required when there are fewer dimensions. Because there is less data, algorithms can be trained more quickly.
- Fewer data necessitates less storage space.
- It removes redundant features and background noise.
- Dimensionality reduction aids in the visualization of data on 2D and 3D graphs.

Dimensionality Reduction Drawbacks

- Some data is lost, which might negatively impact the effectiveness of subsequent training algorithms.
- It has the potential to be computationally demanding.
- Transformed characteristics are often difficult to decipher.
- It makes the independent variables more difficult to comprehend.

3. Can PCA be used to reduce the dimensionality of a nonlinear dataset with many variables?

PCA may be used to dramatically reduce the dimensionality of most datasets, even if they are extremely nonlinear, by removing unnecessary dimensions. However, decreasing dimensionality with PCA will lose too much information if there are no unnecessary dimensions.

4. Is it required to rotate in PCA? If so, why do you think that is? What will happen if the components aren't rotated?

Yes, rotation (orthogonal) is required to account for the training set's maximum variance. If we don't rotate the components, PCA's influence will wane, and we'll have to choose a larger number of components to explain variation in the training set.

5. Is standardization necessary before using PCA?

PCA uses the covariance matrix of the original variables to uncover new directions because the covariance matrix is susceptible to variable standardization. In most cases, standardization provides equal weights to all variables. We obtain false directions when we combine features from various scales. However, if all variables are on the same scale, it is unnecessary to standardize them.

6. Should strongly linked variables be removed before doing PCA?

No, PCA uses the same Principal Component (Eigenvector) to load all strongly associated variables, not distinct ones.

7. What happens if the eigenvalues are almost equal?

PCA will not choose the principle components if all eigenvectors are the same because all principal components will be similar.

8. How can you assess a Dimensionality Reduction Algorithm's performance on your dataset?

A dimensionality reduction technique performs well if it removes many dimensions from a dataset without sacrificing too much information. If you use dimensionality reduction as a preprocessing step before another Machine Learning algorithm (e.g., a Random Forest classifier), you can simply measure the performance of that second algorithm. If dimensionality reduction did not lose too much information, the algorithm should perform well with the original dataset.

The Fourier Transform is a useful image processing method for breaking down an image into sine and cosine components. The picture in the Fourier or frequency domain is represented by the output of the transformation, while the input image represents the spatial domain equivalent.

9. What do you mean when you say "FFT," and why is it necessary?

FFT is an acronym for fast Fourier transform, a DFT computing algorithm. It takes advantage of the twiddle factor's symmetry and periodicity features to drastically reduce its time to compute DFT. As a result, utilizing the FFT technique reduces the number of difficult computations, which is why it is popular.

10. You're well-versed in the DIT Algorithm. Could you tell us more about it?

It calculates the discrete Fourier transform of an N point series and is known as the decimation-in-time algorithm. It divides the sequence into two halves and then combines them to produce the original sequence's DFT. The sequence x(n) is frequently broken down into two smaller subsequences in DIT.

Curse of Dimensionality

When working with high-dimensional data, the "Curse of Dimensionality" refers to a series of issues. The number of attributes/features in a dataset corresponds to the dataset's dimension. High dimensional data contains many properties, usually on a hundred or more. Some of the challenges that come with high-dimensional data appear while analyzing or displaying the data to look for trends, and others show up when training machine learning models.

1. Describe some of the strategies for dimensionality reduction.

The following are some approaches for reducing the dimensionality of a dataset:

- Feature Selection - As we evaluate qualities, we pick or delete them based on their value.
- Feature Extraction - From the current features, we generate a smaller collection of features that summarizes most of the data in our dataset.

2. What are the disadvantages of reducing dimensionality?

Dimensionality reduction has some drawbacks; they include:

- The decrease may take a long time to complete.
- The modified independent variables might be difficult to

comprehend.

- As the number of features is reduced, some information is lost, and the algorithms' performance suffers.

Support Vector Machine (SVM)

The "Support Vector Machine" (SVM) supervised machine learning to solve classification and regression problems. SVMs are especially well-suited to classifying complex but small or medium-sized datasets.

Let's go through several SVM-related interview questions.

1. Could you explain SVM to me?

Support vector machines (SVMs) are supervised machine learning techniques that may be used to solve classification and regression problems. It seeks to categorize data by locating a hyperplane that optimizes the margin between the training data classes. As a result, SVM is a big margin classifier.

Support vector machines are based on the following principle:

For linearly separable patterns, the best hyperplane is extended to patterns that are not linearly separable by original mapping data into new space using modifications of original data (i.e., the kernel trick).

2. In light of SVMs, how would you explain Convex Hull?

We construct a convex hull for classes A and B and draw a perpendicular on the shortest distance between their nearest points.

3. Should you train a model on a training set with millions of instances and hundreds of features using the primal or dual form of the SVM problem?

Because kernelized SVMs may only employ the dual form, this question only relates to linear SVMs. The primal form of the SVM problem has a computational complexity proportional to the number of training examples m. Still, the dual form has a computational complexity proportional to a number between m2 and m3. If there are millions of instances, you should use the primal form instead of the dual form since the dual form is slower.

4. Describe when you want to employ an SVM over a Random Forest Machine Learning method.

The fundamental rationale for using an SVM rather than a linearly separable problem is that the problem may not be linearly separable. We'll have to employ an SVM with a non-linear kernel in such a situation. If you're working in a higher-dimensional space, you can also employ SVMs. SVMs, for example, have been shown to perform better in text classification.

5. Is it possible to use the kernel technique in logistic regression? So, why isn't it implemented in practice?

Logistic regression is more expensive to compute than SVM — O(N3) versus O(N2k), where k is the number of support vectors. The classifier in SVM is defined solely in terms of the support vectors, but the classifier in Logistic Regression is defined over all points, not just the support vectors. This gives SVMs certain inherent speedups (in terms of efficient code-writing) that Logistic Regression struggles to attain.

6. What are the difference between SVM without a kernel and logistic regression?

The only difference is in how they are implemented. SVM is substantially more efficient and comes with excellent optimization tools.

7. Is it possible to utilize any similarity function with SVM?

No, it must comply with Mercer's theorem.

8. Is there any probabilistic output from SVM?

SVMs do not offer probability estimates directly; instead, they are derived through a time-consuming five-fold cross-validation procedure.

Overfitting and Underfitting
Overfitting and Underfitting: An Overview

When the prediction error on both the training and test datasets is large, the model is said to have underfitted, assuming an independent and identically distributed (I.I.d) dataset. This is referred to as model underfitting. Computer learning modelling strategies such as boosting machine learning algorithms, which integrate machine learning models to provide improved predictions using an ensemble of machine-learning model outputs, can tackle the underfitting problem. When underfitting ML models are present, low R-squared values and significant standard errors of estimate in regression analysis, residual plots from machine learning algorithm for linear or logistic regression model output, and so on can all be indicators.

A model is considered to be overfitted when the model accuracy on the training dataset is greater (very high) than the model accuracy on the test dataset. When a machine learning algorithm overfits the training data, the machine learning model may perform well on a small sample of your dataset and provide high accuracy. However, when a machine learning model encounters new input data, its performance suffers dramatically due to overfitting machine learning models to certain patterns in your data set, resulting in memorization of the machine learning model rather than generalization.

Regularization techniques like LASSO (least absolute shrinkage and selection operator) punish large coefficients more strongly than machine learning algorithms like gradient descent. Thus overfitting machine learning models can be avoided or addressed. Overfitting may be detected using a variety of machine learning approaches, such as validation curves and cross-fold plots.

1. What are the many instances in which machine learning models might overfit?

Overfitting of machine learning models can occur in a variety of situations, including the following:

- When a machine learning algorithm uses a considerably bigger training dataset than the testing set and learns patterns in the large input space, the accuracy on a small test set is only marginally improved.
- It occurs when a machine learning algorithm models the training data with too many parameters.
- Suppose the learning algorithm searches a large amount of hypothesis space. Let's figure out what hypothesis space is and what searching hypothesis space is all about. If the learning algorithm used to fit the model has a large number of possible hyperparameters and can be trained using multiple datasets (called training datasets) taken from the same dataset, a large number of models (hypothesis – h(X)) can be fitted on the same data set. Remember that a hypothesis is a target function estimator. As a result, many models may fit the same dataset. This is known as broader hypothesis space. In this case, the learning algorithm can access a broader hypothesis space. Given the broader hypothesis space, the model has a greater chance of overfitting the training dataset.

2. What are the many instances in which machine learning models cause underfitting?

Underfitting of machine learning models can occur in a variety of situations, including the following:

- Underfitting or low-biased machine learning models can occur when the training set contains fewer observations than variables. Because the machine learning algorithm is not complicated enough to represent the data in these circumstances, it cannot identify any link between the input data and the output variable.
- When a machine learning system can't detect a pattern between training and testing set variables, which might happen when dealing with many input variables or a high-dimensional dataset, this might be due to a lack of machine learning model complexity. A scarcity of training observations for pattern learning or a lack of computational power restricts machine learning algorithms' capacity to search for patterns in high-dimensional space, among other factors.

3. What is a Neural Network, and how does it work?

Neural Networks are a simplified version of how people learn, inspired by how neurons in our brains work.

Three network layers make up the most typical Neural Networks:

- There is an input layer.
- A layer that is not visible (this is the most important layer where feature extraction takes place, and adjustments are made to train faster and function better)
- A layer for output

4. What Are the Functions of Activation in a Neural Network?

At its most basic level, an activation function determines whether or not a neuron should activate. Any activation function can take the weighted sum of the inputs and bias as inputs. Activation functions include the step function, Sigmoid, ReLU, Tanh, and Softmax.

5. What is the MLP (Multilayer Perceptron)?

MLPs have an input layer, a hidden layer, and an output layer, just like Neural Networks. It has the same structure as a single layer perceptron with more hidden layers. MLP can identify nonlinear classes, whereas a single layer perceptron can only categorize linear separable classes with binary output (0,1). Each node in the other levels, except the input layer, utilizes a nonlinear activation function. This implies that all nodes and weights are joined together to produce the output based on the input layers, data flowing in, and the activation function. Backpropagation is a supervised learning method used by MLP. The neural network estimates the error with the aid of the cost function in backpropagation. It propagates the mistake backward from the point of origin (adjusts the weights to train the model more accurately).

6. what is Cost Function?

The cost function, sometimes known as "loss" or "error," is a metric used to assess how well your model performs. During backpropagation, it's used to calculate the output layer's error. We feed that mistake backward through the neural network and train the various functions.

7. What is the difference between a Recurrent Neural Network and a Feedforward Neural Network?

The interviewer wants you to respond thoroughly to this deep learning interview question. Signals from the input to the output of a Feedforward Neural Network travel in one direction. The network has no feedback loops and simply evaluates the current input. It is unable to remember prior inputs (e.g., CNN).

The signals of a Recurrent Neural Network go in both directions, resulting in a looped network. It generates a layer's output by combining the present input with previously received inputs and can recall prior data thanks to its internal memory.

8. What can a Recurrent Neural Network (RNN) be used for?

Sentiment analysis, text mining, and picture captioning may benefit from the RNN. Recurrent Neural Networks may also be used to solve problems involving time-series data, such as forecasting stock values over a month or quarter.

5

R LANGUAGE

R is a programming language that you may make as helpful as you wish. It's a tool you have at your disposal that can be used for many things, including statistical analysis, data visualization, data manipulation, predictive modelling, forecast analysis, etc. Google, Facebook, and Twitter are among the firms that use R.

Interview Questions on R

1. What exactly is R?

R is a free and open-source programming language and environment for statistical computation and analysis or data science.

2. What are the various data structures available in R? Explain them in a few words.

These are the data structures that are available in R:

Vector

A vector is a collection of data objects with the same fundamental type, and components are the members of a vector.

Lists

Lists are R objects that include items of various types, such as integers, texts, vectors, or another list.

Matrix

A matrix is a data structure with two dimensions, and vectors of the same length are bound together using matrices. A matrix's elements must all be of the same type (numeric, logical, character).

DataFrame

A data frame, unlike a matrix, is more general in that individual columns might contain various data types (numeric, character, logical, etc.). It is a rectangular list that combines the properties of matrices and lists.

3. What are some of the advantages of R?

Understanding the benefits and drawbacks of various languages and ecosystems is critical, and R is no different. So, what are the benefits of using R?

- It is open-source. For different reasons, this counts as both a benefit and a defect, but being open source means it's publicly available, free to use, and expandable.
- Its ecosystem of packages. As a data scientist, you don't have to spend a lot of time recreating the wheel, thanks to the built-in functions provided by R packages.
- Its statistical and graphical abilities. R's graphing skills,

according to many people, are unrivaled.

4. What are the disadvantages of using R?

You should be aware of the drawbacks of R, just as you should know its benefits.

- Memory and ability to perform. R is often compared to Python as the less powerful language in memory and performance. This is debatable, and many believe it is no longer relevant now that 64-bit systems have taken over the market.
- It's free and open source. Open-source software offers both pros and cons. There is no governing organization in charge of R. Therefore, there is no single point of contact for assistance or quality assurance. This also implies that the R packages aren't always of the best quality.
- Security. Because R was not designed with security in mind, it must rely on third-party resources to fill in the holes.

5. How do you import a CSV file?

It's simple to load a.csv file into R. You have to call the "read.csv()" method and provide it with the file's location .house<-read.csv("C:/Users/John/Desktop/house.csv")

6. What are the various components of graphic grammar?

There are, in general, several components of graphic grammar:

- Facet layer
- Themes layer
- Geometry layer
- Data layer
- Co-ordinate layer
- Aesthetics layer

7. What is Rmarkdown, and how does it work? What's the point of it?

RMarkdown is an R-provided reporting tool. Rmarkdown allows you to produce high-quality reports from your R code. Rmarkdown may produce the following output formats:

- HTML
- PDF
- WORD

8. What is the procedure for installing a package in R?

To install a package in R, do the following command:

install.packages("<package name>")

9. Name a few R programs that can be used for data imputation?

These are some R packages that may be used to input data.

- MICE
- Amelia
- MissForest
- Hmisc
- Mi
- imputeR

10. Can you explain what a confusion matrix is in R?

A confusion matrix can be used to evaluate the model's accuracy. A cross-tabulation of observed and anticipated classes is calculated. The "confusionmatrix()" function from the "caTools" package can be used for this.

11. List some of the functions in the "dplyr" package

The dplyr package includes the following functions:

- Filter
- Select
- Mutate
- Arrange
- Count

12. What would you do if you had to make a new R6 Class?

To begin, we'll need to develop an object template that contains the class's "Data Members" and "Class Functions."

These components make up an R6 object template:

- Private DataMembers
- Name of the class
- Functions of Public Members

13. What do you know about the R package rattle?

Rattle is a popular R-based GUI for data mining. It provides statistical and visual summaries of data, converts data to be easily modeled, creates both unsupervised and supervised machine learning models from the data, visually displays model performance, and scores new datasets for production deployment. One of the most valuable features is that your interactions with the graphical user interface are saved as an R script that can be run in R without using the Rattle interface.

14. What are some R functions which can be used to debug?

The following functions can be used for debugging in R:

- traceback()

- debug()
- browser()
- trace()
- recover()

15. What exactly is a factor variable, and why would you use one?

A factor variable is a categorical variable that accepts numeric or character string values as input. The most important reason to employ a factor variable is that it may be used with great precision in statistical modeling. Another advantage is that they use less memory. To make a factor variable, use the factor() function.

16. In R, what are the three different sorting algorithms?

R's sort() function is used to sort a vector or factor, mentioned and discussed below.

- Radix: This non-comparative sorting method avoids overhead and is usually the most effective. It's a reliable algorithm used to calculate integer vectors and factors.
- Quick Sort: According to R documentation, this function "uses Singleton (1969)'s implementation of Hoare's Quicksort technique and is only accessible when x is numeric (double or integer) and partial is NULL." It isn't regarded as a reliable method.
- Shell: According to the R documentation, this approach "uses Shellsort (an O(n4/3) variation from Sedgewick (1986)."

17. How can R help in data science?

R reduces time-consuming and graphically intense tasks to minutes and keystrokes. In reality, you're unlikely to come across R outside of the world of data science or a related discipline. It's useful for linear and nonlinear modeling, time-series analysis, graphing, grouping, and many other tasks.

Simply put, R was created to manipulate and visualize data. Thus it's only logical that it is used in data science.

18. What is the purpose of the () function in R?

We use a () function to construct simpler code by applying an expression to a data set. Its syntax is as follows:

R Programming Syntax Basics

R is the most widely used language for statistical computing and data analysis, with over 10,000 free packages available in the CRAN library. Like any other programming language, R has a unique syntax that you must learn to utilize all of its robust features.

The R program's syntax

Variables, Comments, and Keywords are the three components of an R program. Variables are used to store data, Comments are used to make code more readable, and Keywords are reserved phrases that the compiler understands.

CSV files in R Programming

CSV files are text files in which each row's values are separated by a delimiter, such as a comma or a tab.

1. Reading a CSV file

The read.csv(...) function in R may read the contents of a CSV file as a data frame. The CSV file must be read in the current working directory, or the directory must be established appropriately in R using the setwd(...) function. The read.csv() method may also read a CSV file via a URL.

2. Using CSV files for querying

The R subset(csv data) function may extract the relevant result from SQL queries on the CSV content. Multiple queries can be run through the function simultaneously, separated by a logical operator. In R, the result is saved as a data frame.

3. Inputting data into a CSV file

The data frame's contents can be saved as a CSV file. The CSV file is saved using the name supplied in R's write.csv(data frame, output CSV name) function in the current working directory.

Confusion Matrix

In R, a confusion matrix is a table that categorizes predictions about actual values. It has two dimensions, one of which will show the anticipated values, and the other will show the actual values.

Each row in the confusion matrix will represent the anticipated values, while the columns represent the actual values. This can also be the case in reverse. The nomenclature underlying the matrixes appears to be sophisticated, even though the matrices themselves are simple. There is always the possibility of being confused regarding the lessons. As a result, the phrase – Confusion matrix – was coined.

The 2×2 matrix in R was visible in the majority of the resources. It's worth noting, however, that you may make a matrix with any number of class values.

A confusion matrix is a value table representing the data points' predicted and actual values. You may use R packages like caret and gmodels and methods like a table() and crosstable() to understand your data better.

Interview Questions on Confusion Matrix

1. What is the purpose of the confusion matrix? Which module do you think you'd use to demonstrate it?

A confusion matrix is one of the simplest ways to summarize the performance of your algorithm in machine learning. Judging the accuracy of a model by looking at the accuracy might be challenging at times because of issues such as unequal distribution. Using a confusion matrix is a better approach to see how good your model is.

Let's start with some definitions.

Classification Accuracy: This is the ratio of the number of right predictions to the number of predictions.

True-positives: These are accurate predictions of real-world events.

False-positives: These are the predictions of actual events that turn incorrect.

True-negatives: True negatives are accurate forecasts of fake events.

False-negatives: False negatives are forecasts of false events that are incorrect.

The confusion matrix has been simplified to a list of true-positives, false-positives, true-negatives, and false-negatives.

2. What is the definition of accuracy?

It's the most basic performance metric, and it's just the ratio of correctly predicted observations to total observations. We may say that it is best if our model is accurate. Yes, accuracy is a valuable statistic, but only when you have symmetric datasets with almost identical false positives and false negatives.

True-Positive + True-Negative / (True-Positive + False-Positive + False-Negative + True-Negative) / (True-Positive + False-Positive + False-Negative + True-Negative) / (True-Positive + False-Positive + False-Negative + True-Negative

3. What is the definition of precision?

It's also referred to as the positive predictive value. In your predictive model, precision is the number of right positives as compared to the overall number of positives it forecasts.

True-Positives / (True-Positives + False-Positives) Precision = True-Positives / (True-Positives + False-Positives). True-Positives / Total Predicted Positives = Precision

It's the number of correctly predicted positive items divided by the total number of correctly predicted positive elements. Precision may be defined as a metric of exactness, quality, or correctness.

Exceptional accuracy: This indicates that most, if not all, of the good outcomes you predicted, are right.

4. What is the definition of recall?

Recall that we may also refer to this as sensitivity or true-positive rate. The model predicts many positives compared to our data's actual number of positives.

True-Positives/(True-Positives + False-Positives) = Recall

True-Positives / Total Actual Positives = Recall

A recall measures completeness. Our model had a high recall, implying it categorized most or all positive aspects as positive.

Random Forest in R

1. What is your definition of Random Forest?

Random Forest is a form of ensemble learning approach for classification, regression, and other tasks related to Random Forests. Random Forests works by training a large number of decision trees simultaneously, and this is accomplished by averaging many decision trees from various portions of the same training set.

2. What are the outputs of Random Forests for Classification and Regression problems?

- Classification: The Random Forest's output is chosen by the most trees.
- Regression: The mean or average forecast of the various trees is the Random Forest's output.

3. What do Ensemble Methods entail?

Ensemble techniques are a machine learning methodology that integrates numerous base models to create a single best-fit prediction model. Random Forest are a form of ensemble method.

However, there is a law of decreasing returns in ensemble formation. The number of component classifiers in an ensemble significantly influences the accuracy of the prediction.

4. What are some Random Forest hyperparameters?

Hyperparameters in Random Forest include:

- The forest's total number of decision trees.
- The number of characteristics that each tree considers while splitting a node.
- The individual tree's maximum depth.

- The minimum number of samples to divide at an internal node.
- The number of leaf nodes at its maximum.
- The total number of random characteristics
- The bootstrapped dataset's size.

5. How would you determine the Bootstrapped Dataset's ideal size?

Even though the size of the bootstrapped dataset is different, the datasets will be different since the observations are sampled with replacements.

As a result, the training data may be used in its entirety. The best thing to do most of the time is ignoring this hyperparameter.

6. Is it necessary to prune Random Forest? Why do you think that is?

Pruning is a data compression method used in machine learning and search algorithms to minimize the size of decision trees by deleting non-critical and redundant elements of the tree.

Because it does not over-fit like a single decision tree, Random Forest typically does not require pruning. This occurs when the trees are bootstrapped and numerous random trees employ random characteristics, resulting in robust individual trees not associated with one another.

7. Is it required to use Random Forest with Cross-Validation?

A random forest's OOB is comparable to Cross-Validation, and as a result, cross-validation is not required. By default, random forest uses 2/3 of the data for training, the remainder for testing in regression, and about 70% for training and testing in classification. Because the variable selection is randomized during each tree split, it is not prone to overfitting like other models.

8. What is the relationship between a Random Forest and Decision Trees?

Random forest is an ensemble learning approach that uses many decision trees to learn. A random forest may be used for classification and regression, and random forest outperforms decision trees and does not have the same tendency to overfit the data.

Overfitting occurs when a decision tree trained on a given dataset becomes too deep. Decision trees may be trained on multiple subsets of the training information to generate a random forest, and then the different decision trees can be averaged to reduce variation.

9. Is Random Forest an Ensemble Algorithm?

Yes, Random Forest is a tree-based ensemble technique that relies on a set of random variables for each tree. Bagging is used as the ensemble approach, while decision tree is used as the individual model in Random Forest.

Random forests can be used for classification, regression, and other tasks in which a large number of decision trees are built at the same time. The random forest's output is the class most trees choose for classification tasks.

The mean or average forecast of the individual tresses is returned for regression tasks. Decision trees tend to overfit their training set, corrected by random forests.

K-MEANS Clustering
1. What are some examples of k-Means Clustering applications?

The following are some examples of k-means clustering applications:

- Document classification: Based on tags, subjects, and the document's substance, k-means may group documents into numerous groups.

- Insurance fraud detection: It is feasible to identify new claims based on their closeness to clusters that signal fraudulent tendencies using previous data on fraudulent claims.

- Criminals who use cyber-profiling: This is the practice of gathering data from people and groups to find significant correlations. Cyber profiling is based on criminal profiles, which offer information to the investigation division to categorize the sorts of criminals present at the crime scene.

2. How can you tell the difference between KNN and K-means clustering?

The K-nearest neighbor algorithm is a supervised classification method known as KNN. This means categorizing an unlabeled data point, requiring labeled data. It tries to categorize a data point in the feature space based on its closeness to other K-data points.

K-means Clustering is a method for unsupervised classification. It merely needs a set of unlabeled points and a K-point threshold to collect and group data into K clusters.

3. What is k-Means Clustering?

K-means Clustering is a vector quantization approach that divides a set of n observations into k clusters, with each observation belonging to the cluster with the closest mean. Within-cluster variances are minimized using k-means clustering.

Within-cluster-variance is an easy-to-understand compactness metric. Essentially, the goal is to split the data set into k divisions in the most compact way possible.

4. What is the Uniform Effect produced by k-Means Clustering?

The Uniform Effect refers to the tendency of k-means clustering to create clusters of uniform size. Even if the data behaves differently, uniform sizes ensure that the clusters have about the same number of observations.

5. What are some k-Means Clustering Stopping Criteria?

The following are some of the most common reasons for stopping:

- Convergence. There are no more modifications; the points remain in the same cluster.
- The number of iterations that can be done. The method will be terminated after the maximum number of iterations has been reached. This is done to keep the algorithm's execution time to a minimum.
- Variance hasn't increased by at least x%.
- The variance did not increase by more than x times the starting variance.
- MiniBatch k-means will not converge, so one of the other criteria is required. The number of iterations is the most common.

6. Why does the Euclidean Distance metric dominate in k-Means Clustering?

The construction of k-means is not reliant on distances, and Within-cluster variance is decreased using K-means. When you examine the variance definition, you'll notice that it's the sum of squared Euclidean distances from the center.

The goal of k-means is to reduce squared errors. There is no such thing as "distance" in this case. Pairwise distances between data points are not explicitly used in the k-means process. It entails assigning points to the nearest centroid over and over again, based on the Euclidean distance between data points and a centroid.

Euclidean geometry is the origin of the term "centroid." In Euclidean space, it is a multivariate mean. Euclidean distances are the subject of Euclidean space. In most cases, non-Euclidean distances will not cross Euclidean space, which is why K-Means is only used for Euclidean distances.

Using arbitrary distances is incorrect because k-means may stop converging with other distance functions.

6

SQL

Let's delve straight into some SQL interview questions.

1. What exactly is SQL?

SQL is an acronym for the structured query language. It is a database management system that allows you to access and manipulate data. In 1986, the American National Standards Institute (ANSI) approved SQL as a standard.

2. What Can SQL do for you?

- SQL is capable of running queries against a database.
- SQL may be used to get information from a database.
- SQL may be used to create new records in a database.
- SQL may be used to update data in a database.
- SQL can delete records from a database.
- SQL can build new databases.
- SQL can create new tables in a database.
- In a database, SQL may build stored procedures.
- In a database, SQL may be used to generate views.

- Permissions can be established on tables, methods, and views in SQL.

Interview Questions for Basic SQL

1. How do you distinguish between SQL and MySQL?

SQL is a standard language based on English. MySQL is a relational database management system (RDBMS). SQL is the foundation of a relational database, and it is used to retrieve and manage data. MySQL is a relational database management system (RDMS), similar to SQL Server and Informix.

2. What are the various SQL subsets?

- Data Definition Language (DDL) lets you do things like CREATE, ALTER, and DELETE items on the database.
- Data Manipulation Language (DML) allows you to alter and access data. It aids in inserting, updating, deleting, and retrieving data from a database.
- Data Control Language (DCL) allows you to manage database access, grant and revoke access permissions.

3. What do you mean by database management system (DBMS)? What are the many sorts of it?

A Database Management System (DBMS) is a software program that captures and analyzes data through interacting with the user, applications, and the database itself. A database is a collection of data that is organized.

A database management system (DBMS) allows users to interface. The database's data may be edited, retrieved, and destroyed, and it can be of any type, including strings, integers, and pictures.

There are two types of database management systems (DBMS):

- Relational Database Management System (RDBMS): Information is organized into relationships (tables). MySQL is a good example.
- Non-Relational Database Management System: This system has no relations, tuples, or attributes. A good example is MongoDB.

4. In SQL, how do you define a table and a field?

A table is a logically organized collection of data in rows and columns. The number of columns in a table is referred to as a field. Consider the following scenario:

Fields: Student ID, Student Name, and Student Marks

5. How do we define joins in SQL?

A join clause joins rows from two or more tables based on a common column. It's used to join two tables together or derive data from them. As seen below, there are four different types of joins:

- Inner join: The most frequent join in SQL is the inner join. It's used to get all the rows from various tables that satisfy the joining requirement.
- Full Join: When there is a match in any table, a full join returns all the records. As a result, all rows from the left-hand side table and all rows from the right-hand side table are returned.
- Right Join: In SQL, a "right join" returns all rows from the right table but only matches records from the left table when the join condition is met.
- Left Join: In SQL, a left join returns all of the data from the left table, but only the matching rows from the right table when the join condition is met.

6. What is the difference between the SQL data types CHAR and VARCHAR2?

Both Char and Varchar2 are used for character strings. However, Varchar2 is used for variable-length strings, and Char is used for fixed-length strings. For instance, char (10) can only hold 10 characters and cannot store a string of any other length, but varchar2 (10) may store any length, i.e. 6, 8, 2.

7. What are constraints?

In SQL, constraints are used to establish the table's data type limit. It may be supplied when the table statement is created or changed. The following are some examples of constraints:

- UNIQUE
- NOT NULL
- FOREIGN KEY
- DEFAULT
- CHECK
- PRIMARY KEY

8. What is a foreign key?

A foreign key ensures referential integrity by connecting the data in two tables. The foreign key as defined in the child table references the primary key in the parent table. The foreign key constraint obstructs actions to terminate links between the child and parent tables.

9. What is"data integrity"?

Data integrity refers to the consistency and correctness of data kept in a database. It also specifies integrity constraints, which are used to impose business rules on data when input into an application or database.

10. What is the difference between a clustered and a non-clustered index?

The following are the distinctions between a clustered and non-clustered index in SQL:

- Clustered indexes are utilized for quicker data retrieval from databases, whereas reading from non-clustered indexes takes longer.
- A clustered index changes the way records are stored in a database by sorting rows by the clustered index column. A non-clustered index does not change the way records are stored but instead creates a separate object within a table that points back to the original table rows after searching.

There can only be one clustered index per table, although there can be numerous non clustered indexes.

11. How would you write a SQL query to show the current date?

A built-in method in SQL called GetDate() returns the current timestamp/date.

12. What exactly do you mean when you say "query optimization"?

Query optimization is the step in which a plan for evaluating a query that has the lowest projected cost is identified.

The following are some of the benefits of query optimization:

- The result is delivered more quickly.
- In less time, a higher number of queries may be run.
- Reduces the complexity of time and space

13. What is "denormalization"?

Denormalization is a technique for retrieving data from higher to lower levels of a database. It aids database administrators in improving the overall performance of the infrastructure by introducing redundancy into a table. It incorporates database queries that merge data from many tables into a single table to add redundant data to a table.

14. What are the differences between entities and relationships?

Entities are real-world people, places, and things whose data may be kept in a database. Tables are used to contain information about a single type of object. A customer table, for example, is used to hold customer information in a bank database. Each client's information is stored in the customer database as a collection of characteristics (columns inside the table).

Relationships are connections or connections between things that have something in common. The customer name, for example, is linked to the customer account number and contact information, which may be stored in the same database. There may also be connections between different tables (for example, customer to accounts).

15. What is an index?

An index is a performance optimization strategy for retrieving records from a table quickly. Because an index makes an entry for each value, retrieving data is faster.

16. Describe the various types of indexes in SQL.

In SQL, there are three types of indexes:

- Unique Index: If the column is unique indexed, this index prevents duplicate values in the field. A unique index can be applied automatically if the main key is provided.
- Clustered Index: This index reorders the table's physical order and searches based on key values. There can only be one

clustered index per table.
- Non-Clustered Index: Non-clustered indexes do not change the physical order of the database and keep the data in a logical order. There might be a lot of nonclustered indexes in a table.

17. What is normalization, and what are its benefits?

The practice of structuring data in SQL to prevent duplication and redundancy is known as normalization. The following are some of the benefits:

- Improved database management
- Tables with smaller rows are added to the mix.
- Efficient data access
- Greater queries flexibility
- Locate the information quickly.
- Security is easier to implement.
- Allows for easy customization/customization
- Data duplication and redundancy are reduced.
- More compact database.
- Ensure that data is consistent after it has been modified.

18. Describe the various forms of normalization.

There are several levels of normalization to choose from. These are referred to as normal forms. Each subsequent normal form is dependent on the one before it. In most cases, the first three normal forms are sufficient.

- First Normal Form (1NF) – There are no repeating groups in between rows
- Second Normal Form (2NF) – Every non-key (supporting) column value relies on the primary key.
- Third Normal Form (3NF) – Dependent solely on the primary

key and no other non-key (supporting) column value.

19. In a database, what is the ACID property?

Atomicity, Consistency, Isolation, and Durability (ACID) is used to verify that data transactions in a database system are processed reliably.

- Atomicity: Atomicity relates to completed or failed transactions. A transaction refers to a single logical data operation. It means that if one portion of a transaction fails, the full transaction fails as well, leaving the database state unaltered.
- Consistency: Consistency guarantees that the data adheres to all validation standards. In basic terms, your transaction never leaves the database before it has completed its state.
- Isolation: The main purpose of isolation is concurrency control.
- Durability: Durability refers to the fact that once a transaction has been committed, it will occur regardless of what happens in the meantime, such as a power outage, a crash, or any other type of mistake.

20. What is "Trigger" in SQL?

Triggers are a stored procedure in SQL that is configured to execute automatically in situ or after data changes. When an insert, update, or other query is run against a specified table, it allows you to run a batch of code.

21. What are the different types of SQL operators?

In SQL, there are three operators available:

- Logical Operators
- Arithmetic Operators
- Comparison Operators

22. Do NULL values have the same meaning as zero or a blank space?

A null value is not confused with a value of zero or a blank space. A null value denotes an unavailable, unknown, assigned, or not applicable value, whereas a zero denotes a number and a blank space denotes a character.

23. What is the difference between a natural join and a cross join?

The natural join is dependent on all columns in both tables having the same name and data types, whereas the cross join creates the cross product or Cartesian product of two tables.

24. What is a subquery in SQL?

A subquery is a query defined inside another query to get data or information from the database. The outer query of a subquery is referred to as the main query. In contrast, the inner query is referred to as the subquery. Subqueries are always processed first, and the subquery's result is then passed on to the main query. It may be nested within any query, including SELECT, UPDATE, and OTHER. Any comparison operators, such as >, or =, can be used in a subquery.

25. What are the various forms of subqueries?

Correlated and Non-Correlated subqueries are the two forms of the subquery.

- Correlated subqueries: These queries pick data from a table that the outer query refers to. It is not considered an independent query because it refers to another table and column.
- Non-Correlated subquery: This query is a stand-alone query in which a subquery's output is used to replace the main query results.

DBMS and RDBMS
1. What is a database management system (DBMS), and what is its purpose? Use examples to explain RDBMS.

The database management system, or DBMS, is a collection of applications or programs that allow users to construct and maintain databases. A database management system (DBMS) offers a tool or interface for executing different database activities such as adding, removing, updating, etc. It is software that allows data to be stored more compactly and securely than a file-based system. A database management system (DBMS) assists a user in overcoming issues such as data inconsistency, data redundancy, and other issues in a database, making it more comfortable and organized to use.

Examples of prominent DBMS systems are file systems, XML, the Windows Registry, and other DBMS systems.

RDBMS stands for Relational Database Management System, and it was first introduced in the 1970s to make it easier to access and store data than DBMS. In contrast to DBMS, which stores data as files, RDBMS stores data as tables. Unlike DBMS, storing data in rows and columns makes it easier to locate specific values in the database and more efficient.

MySQL, Oracle DB, are good examples of RDBMS systems.

2. What is a database?

A database is a collection of well-organized, consistent, and logical data and can be readily updated, accessed, and controlled. Most databases are made up of tables or objects (everything generated with the create command is a database object) that include entries and fields. A tuple or row represents a single entry in a table. The main components of data storage are attributes and columns, which carry information about a specific element of the database. A database management system

(DBMS) pulls data from a database using queries submitted by the user.

3. What drawbacks of traditional file-based systems make a database management system (DBS) a superior option?

The lack of indexing in a typical file-based system leaves us little choice but to scan the whole page, making content access time-consuming and sluggish. The other issue is redundancy and inconsistency, as files often include duplicate and redundant data, and updating one causes all of them to become inconsistent. Traditional file-based systems make it more difficult to access data since it is disorganized.

Another drawback is the absence of concurrency management, which causes one action to lock the entire page, unlike DBMS, which allows several operations to operate on the same file simultaneously.

Integrity checking, data isolation, atomicity, security, and other difficulties with traditional file-based systems have all been addressed by DBMSs.

4. Describe some of the benefits of a database management system (DBS).

The following are some of the benefits of employing a database management system (DBS).

- Data Sharing: Data from a single database may be shared by several users simultaneously. End-users can also respond fast to changes in the database environment because of this sharing.
- Integrity restrictions: The presence of such limitations allows for the ordered and refined storage of data.
- Controlling database redundancy: Provides a means for integrating all data in a single database, eliminating redundancy in a database.
- Data Independence: This allows you to change the data structure

without affecting the composition of any of the application programs that are currently running.

- Provides backup and recovery facility: It may be configured to automatically generate a backup of the data and restore the data in a database when needed.
- Data Security: A database management system (DBMS) provides the capabilities needed to make data storage and transmission more dependable and secure. Some common technologies used to safeguard data in a DBMS include authentication (the act of granting restricted access to a user) and encryption (encrypting sensitive data such as OTP, credit card information, and so on).

5. Describe the different DBMS languages.

The following are some of the DBMS languages:

- DDL (Data Definition Language) is a language that includes commands for defining databases.
- CREATE, ALTER, DROP, TRUNCATE, RENAME, and so on.
- DML (Data Manipulation Language) is a set of instructions that may alter data in a database. SELECT, UPDATE, INSERT, DELETE, and so on.
- DCL (Data Control Language): It offers instructions for dealing with the database system's user permissions and controls.
- GRANT and REVOKE, for example.
- TCL (Transaction Control Language) is a programming language that offers instructions for dealing with database transactions. COMMIT, ROLLBACK, and SAVEPOINT are a few examples.

6. What does it mean to have ACID qualities in a database management system (DBMS)?

In a database management system, ACID stands for Atomicity, Consistency, Isolation, and Durability. These features enable a safe and secure exchange of data among different users.

- Atomicity: This attribute supports the notion of either running the whole query or doing nothing at all, which means that if a database update occurs, it should either be reflected across the entire database or not at all.
- Consistency: This feature guarantees that data is consistent before and after a transaction in a database.
- Isolation: This characteristic assures that each transaction is separate from the others, and this suggests that the status of one ongoing transaction has no bearing on the condition.
- Durability: This attribute guarantees that data is not destroyed in the event of a system failure or restart and that it is available in the same condition as before the failure or restart.

7. Are NULL values in a database the same as blank space or zero?

No, a null value is different from zero and blank space. It denotes a value that is assigned, unknown, unavailable, or not applicable, as opposed to blank space, which denotes a character, and zero, which denotes a number.

For instance, a null value in the "number of courses" taken by a student indicates that the value is unknown, but a value of 0 indicates that the student has not taken any courses.

8. What does Data Warehousing mean?

Data warehousing is the process of gathering, extracting, processing, and importing data from numerous sources and storing it in a single database. A data warehouse may be considered a central repository for data analytics that receives data from transactional systems and other relational databases. A data warehouse is a collection of historical data from an organization that aids in decision-making.

9. Describe the various data abstraction layers in a database management system (DBMS).

Data abstraction is the process of concealing extraneous elements from consumers. There are three degrees of data abstraction:

- Physical Level: This is the lowest level, and the database management system maintains it. The contents of this level are often concealed from system admins, developers, and users, and it comprises data storage descriptions.
- Conceptual or logical level: Developers and system administrators operate at the conceptual or logical level, which specifies what data is kept in the database and how the data points are related.
- External or View level: This level only depicts a portion of the database and keeps the table structure and actual storage specifics hidden from users. The result of a query is an example of data abstraction at the View level. A view is a virtual table formed by choosing fields from multiple database tables.

10. What does an entity-relationship (E-R) model mean? Define an entity, entity type, and entity set in a database management system.

A diagrammatic approach to database architecture in which real-world things are represented as entities and connections between them are indicated is known as an entity-relationship model.

- Entity: A real-world object with attributes that indicate the item's qualities is defined as an entity. A student, an employee, or a teacher, for example, symbolizes an entity.

- Entity Type: This is a group of entities with the same properties. An entity type is represented by one or more linked tables in a database. Entity type or attributes may be thought of as a trait that distinguishes the entity from others. A student, for example, is an entity with properties such as student id, student name, and so on.

- Entity Set: An entity set is a collection of all the entities in a database that belongs to a given entity type. An entity set, for example, is a collection of all students, employees, teachers, and other individuals.

11. What is the difference between intension and extension in a database?

The main distinction between intension and extension in a database is as follows:

- Intension: Intension, also known as database schema, describes the database's description. It is specified throughout the database's construction and typically remains unmodified.

- Extension, on the other hand, is a measurement of the number of tuples in a database at any particular moment in time. The snapshot of a database is also known as the extension of a database. The value of the extension changes when tuples are created, modified, or deleted in the database.

12. Describe the differences between the DELETE and TRUNCATE commands in a database management system.

DELETE command: this command is used to delete rows from a table based on the WHERE clause's condition.

- It just deletes the rows that the WHERE clause specifies.
- If necessary, it can be rolled back.

- It keeps a record to lock the table row before removing it, making it sluggish.
- The TRUNCATE command is used to delete all data from a table in a database. Consequently, making it similar to a DELETE command without a WHERE clause.
- It deletes all of the data from a database table.
- It may be rolled back if necessary. (Truncate can be rolled back, but it's hard and can result in data loss depending on the database version.)
- It doesn't keep a log and deletes the entire table at once, so it's quick.

13. Define lock. Explain the significant differences between a shared lock and an exclusive lock in a database transaction.

A database lock is a method that prevents two or more database users from updating the same piece of data at the same time. When a single database user or session obtains a lock, no other database user or session may edit the data until the lock is released.

- Shared lock: A shared lock is necessary for reading a data item, and in a shared lock, many transactions can hold a lock on the same data item. A shared lock allows many transactions to read the data items.
- Exclusive lock: A lock on any transaction that will conduct a write operation is an exclusive lock. This form of lock avoids inconsistency in the database by allowing only one transaction at a time.

14. What do normalization and denormalization mean?

Normalization is breaking up data into numerous tables to reduce duplication. Normalization allows for more efficient storage space and makes maintaining database integrity.

Denormalization is the reversal of normalization, in which tables that have been normalized are combined into a single table to speed up data retrieval. By reversing the normalization, the JOIN operation allows us to produce a denormalized data representation.

RDBMS

1. What are the various characteristics of a relational database management system (RDBMS)?

- Name: Each relation should have a distinct name from all other relations in a relational database.
- Attributes: An attribute is a name given to each column in a relation.
- Tuples: Each row in a relation is referred to as a tuple. A tuple is a container for a set of attribute values.

2. What is the E-R Model, and how does it work?

The E-R model stands for Entity-Relationship. The E-R model is based on a real-world environment that consists of entities and related objects. A set of characteristics is used to represent entities in a database.

3. What does an object-oriented model entail?

The object-oriented paradigm is built on the concept of collections of items. Values are saved in instance variables within an object and stored. Classes are made up of objects with the same values and use the same methods.

4. What are the three different degrees of data abstraction?

- Physical level: This is the most fundamental level of abstraction, describing how data is stored.
- Logical level: The logical level of abstraction explains the types of data recorded in a database and their relationships.

- View level: This is the most abstract level, and it describes the entire database.

5. What are the differences between Codd's 12 Relational Database Rules?

Edgar F. Codd presented a set of thirteen rules (numbered zero to twelve) that he called Codd's 12 rules.

Codd's rules are as follows:

- Rule 0: The system must meet Relational, Database, and Management Systems requirements.
- Rule 1: The information rule: Every piece of data in the database must be represented uniquely, most notably name values in column locations inside a distinct table row.
- Rule 2: The second rule is the assured access rule, which states that all data must be ingressive. Every scalar value in the database must be correctly/logically addressable.
- Rule 3: Null values must be treated consistently: The DBMS must allow each tuple to be null.
- Rule 4: Based on the relational paradigm, an active online catalog (database structure): The system must provide an online, relational, or other structure that is ingressive to authorized users via frequent queries.
- Rule 5: The sublanguage of complete data: The system must support at least one relational language that meets the following criteria:
  1. That has a linear syntax
  2. That can be utilized interactively as well as within application applications.
  3. Data definition (DDL), data manipulation (DML), security and integrity restrictions, and transaction management activities are all supported (begin, commit,

and roll back).

- Rule 6: The view update rule: The system must upgrade any views that theoretically improve.

- Rule 7: Insert, update, and delete at the highest level: The system must support insert, update, and remove operators at the highest level.

- Rule 8: Physical data independence: Changing the physical level (how data is stored, for example, using arrays or linked lists) should not change the application.

- Rule 9: Logical data independence: Changing the logical level (tables, columns, rows, and so on) should not need changing the application.

- Rule 10: Integrity independence: Each application program's integrity restrictions must be recognized and kept separately in the catalog.

- Rule 11: Distribution independence: Users should not see how pieces of a database are distributed to multiple sites.

- Rule 12: The nonsubversion rule: If a low-level (i.e., records) interface is provided, that interface cannot be used to subvert the system.

6. What is the definition of normalization? What, therefore, explains the various normalizing forms?

Database normalization is a method of structuring data to reduce data redundancy. As a result, data consistency is ensured. Data redundancy has drawbacks, including wasted disk space, data inconsistency, and delayed DML (Data Manipulation Language) searches. Normalization forms include 1NF, 2NF, 3NF, BCNF, 4NF, 5NF, ONF, and DKNF.

- 1.1NF: Each column's data should contain atomic number multiple values separated by a comma. There are no recurring column groupings in the table, and the main key is used to identify each entry individually.

- 2.2NF: – The table should satisfy all of 1NF's requirements, and redundant data should be moved to a separate table. Furthermore, it uses foreign keys to construct a link between these tables.
- 3.3NF: A 3NF table must meet all of the 1NF and 2NF requirements. There are no characteristics in 3NF that are partially reliant on the main key.

7. What are primary key, a foreign key, a candidate key, and a super key?

- The main key is the key that prevents duplicate and null values from being stored. A primary key can be specified at the column or table level, and per table, only one primary key is permitted.
- Foreign key: a foreign key only admits values from the linked column, and it accepts null or duplicate values. It can be classified as either a column or a table level, and it can point to a column in a unique/primary key table.
- Candidate Key: A Candidate key is the smallest super key; no subset of Candidate key qualities may be used as a super key.
- A super key: is a collection of related schema characteristics on which all other schema elements are partially reliant. The values of super key attributes cannot be identical in any two rows.

8. What are the various types of indexes?

The following are examples of indexes:

- Clustered index: This is where data is physically stored on the hard drive. As a result, a database table can only have one clustered index.
- Non-clustered index: This index type does not define physical data but defines logical ordering. B-Tree or B+ trees are

commonly used for this purpose.

9. What are the benefits of a relational database management system (RDBMS)?

Controlling Redundancy is the answer.

- Integrity is something that can be enforced.
- It is possible to prevent inconsistency.
- It's possible to share data.
- Standards are enforceable.

10. What are some RDBMS subsystems?

RDBMS subsystems are Language processing, Input-output, security, storage management, distribution control, logging and recovery, transaction control, and memory management.

11. What is Buffer Manager, and how does it work?

The Buffer Manager collects data from disk storage and chooses what data should be stored in cache memory for speedier processing.

MYSQL

MySQL is a relational database management system that is free and open-source (RDBMS). It works both on the web and on the server. MySQL is a fast, dependable, and simple database, and it's a free and open-source program. MySQL is a database management system that runs on many systems and employs standard SQL. It's a SQL database management system that's multithreaded and multi-user.

Tables are used to store information in a MySQL database. A table is a set of columns and rows that hold linked information.

MySQL includes standalone clients that allow users to communicate directly with a MySQL database using SQL. Still, MySQL is more common to be used in conjunction with other programs to create applications that require relational database functionality.

Over 11 million people use MySQL.

Basic MYSQL Interview Questions

1. What exactly is MySQL?

MySQL is a scalable web server database management system, and it can expand with the website. MySQL is by far the most widely used open-source SQL database management system, developed by Oracle Corporation.

2. What are a few of the benefits of MySQL?

- MySQL is a flexible database that operates on any operating system.
- MySQL is focused on performance.
- SQL at the Enterprise Level MySQL had been deficient in sophisticated functionality like subqueries, views, and stored procedures for quite some time.
- Indexing and Searching of Full-Text Documents
- Query Caching: This significantly improves MySQL's performance.
- Replication: A MySQL server may be copied on another, with many benefits.
- Security and configuration

3. What exactly do you mean when you say "databases"?

A database is a structured collection of data saved in a computer system and organized to be found quickly. Information may be quickly found via databases.

4. What does SQL stand for in MySQL?

SQL stands for Structured Query Language in MySQL. Other databases, such as Oracle and Microsoft SQL Server, also employ this language. To submit queries from a database, use instructions like the ones below:

It's worth noting that SQL doesn't care about the case. However, writing SQL keywords in CAPS and other names and variables in a small case is a good practice.

5. What is a MySQL database made out of?

A MySQL database comprises one or more tables, each with its own set of entries or rows. The data is included in numerous columns or fields inside these rows.

6. What are your options for interacting with MySQL?

You may communicate with MySQL in three different ways:

- Via a web interface
- Using a command line
- Through a programming language

7. What are MySQL Database Queries, and How do I use them?

An inquiry is a request or a precise question. A database may be queried for specific information, and a record returned.

8. In MySQL, what is a BLOB?

The abbreviation BLOB denotes a big binary object, and its purpose is to store a changeable amount of information.

There are four different kinds of BLOBs:

- TINYBLOB
- MEDIUMBLOB
- BLOB
- LONGBLOB

A BLOB may store a lot of information. Documents, photos, and even films are examples. If necessary, you may save the whole manuscript as a BLOB file.

9. What is the procedure for adding users to MySQL?

By executing the CREATE command and giving the required credentials, you may create a User. Consider the following scenario:

CREATE USER 'testuser' WITH' sample password' AS IDENTIFIER.

10. What exactly are MySQL's "Views"?

A view in MySQL is a collection of rows that are returned when a certain query is run. A 'virtual table' is another name for this. Views make it simple to find out how to make a query available via an alias.

Views provide the following advantages:

- Security
- Simplicity
- Maintainability

11. Define MySQL Triggers?

A trigger is a job that runs in reaction to a predefined database event, such as adding a new record to a table. This event entails entering, altering, or removing table data, and the action might take place before or immediately after any such event.

Triggers serve a variety of functions, including:

- Validation
- Audit Trails
- Referential integrity enforcement

12. In MySQL, how many triggers are possible?

There are six triggers that may be used in the MySQL database:

- After Insert
- Before Insert
- Before Delete
- Before Update
- After Update
- After Delete

13. What exactly is a MySQL server?

The server, mySQLd, is the heart of a MySQL installation; it handles all database and table management.

14. What are MySQL's client and utility programs?

To interact with the server, you can use several MySQL applications. Some of the most significant administrative responsibilities are outlined below:

- mySQL—An interactive application for sending SQL commands to a server and viewing the results. MySQL may also run batch scripts (text files containing SQL statements).
- mySQLadmin—An administrative application for activities like shutting down the server, reviewing its setup, and monitoring its status if it doesn't appear to be working properly.
- mySQLdump—A utility for backing up and transferring databases from one server to another.
- mySQLcheck and myisamchk—Programs that let you check,

analyze, and optimize tables, as well as repair them if they become damaged. MyISAM tables and, to a lesser extent, tables for other storage engines are supported by mySQLcheck. Only MyISAM tables should be used with myisamchk.

14. What are the different types of MySQL relationships?

In MySQL, there are three types of relationships:

- One-o-One: When two things have a one-to-one relationship, they are usually included as columns in the same table.
- One-to-Many: When one row in one database is linked to many rows in another table, this is known as a one-to-many (or many-to-one) connection.
- Many-to-Many: Many rows in one table are connected to many rows in another table in a many-to-many connection. Add a third table with the same key column as the other tables 29 to establish this link.

15. What is MySQL Scaling?

In MySQL, scaling capacity refers to the system's ability to manage demand, and it's helpful to consider load from a variety of perspectives, including:

- Quantity of information
- Amount of users
- Size of related datasets
- User activity

16. What is SQL Sharding?

Sharding divides huge tables into smaller portions (called shards) distributed across different servers. The benefit of sharding is that searches, maintenance, and other operations are quicker because the sharded database is typically much smaller than the original.

Unique Constraints

The rule that states that the values of a key are valid only if they are unique is known as the unique constraint. A unique key has just one set of values, and a unique index is utilized to apply a unique restriction. During the execution of INSERT and UPDATE commands, the database manager utilizes the unique index to guarantee that the values of the key are unique.

There are two kinds of Unique constraints:

A CREATE TABLE or ALTER TABLE command can specify a unique key as a primary key. There can't be more than one main key in a base table. A CHECK constraint will be introduced automatically to enforce the requirement that NULL values are not permitted in the primary key fields. The main index is a unique index on a primary key.

The UNIQUE clause of the CREATE TABLE or ALTER TABLE statement may be used to establish unique keys. There can be many sets of UNIQUE keys in a base table, and there are no restrictions on the number of null values that can be used.

The parent key is a unique key referenced by the foreign key of a referential constraint. The main key or a UNIQUE key is a parent key, and the default parent key is its main key when a base table is designated as a parent in a referential constraint.

When a unique constraint is defined, the unique index used to enforce it is constructed implicitly. Alternatively, the CREATE UNIQUE INDEX statement can be used to define it.

1. What are constraints?

A constraint is an attribute of a table column that conducts data validation. Constraints help to ensure data integrity by prohibiting the entry of incorrect data.

2. What do you mean when you say "data integrity"?

The consistency and correctness of data kept in a database are data integrity.

3. Is it possible to add constraints to a table that already contains data?

Yes, but it also depends on the data. For example, if a column contains null values and adds a not-null constraint, you must first replace all null values with some values.

4. Can a table have more than one primary key?

No table can only have one primary key

5. What is the definition of a foreign key?

In one table, an FK refers to a PK in another. It prohibits any operations that might break the linkages between tables and the data values they represent. FKs are used to ensure that referential integrity is maintained.

6. What is the difference between primary and unique key constraints?

- A null value will be allowed if the constraint is unique. A unique constraint will allow just one null value if a field is nullable.
- SQL Server allows for several unique constraints per table, but MySQL only allows for a single primary key.

7. Is it possible to use Unique key restrictions across multiple columns?

Yes! Unique key constraints can be imposed on a composite of many fields to assure record uniqueness.

Example: City + State in the StateList table

8. When you add a unique key constraint, which index does the database construct by default?

A nonclustered index is constructed when you add a unique key constraint.

9. What does it mean when you say "default constraints"?

When no value is supplied in the Insert or Update statement, a default constraint inserts a value in the column.

10. What kinds of data integrity are there?

There are three types of integrity in relational databases.

- Entity Integrity (unique constraints, primary key)
- Domain Integrity (check constraints, data type)

Clustered and Non-Clustered Indexes
1. What exactly is an index?

An index is a database object that the SQL server uses to improve query performance by allowing query access to rows in the data table. We can save time and increase the speed of database queries and applications by employing indexes.

When constructing an index on a column, SQL Server creates a second index table. When a user tries to obtain data from an existing table that relies on the index table, SQL Server goes straight to the table and quickly retrieves the data.

250 indexes may be used in a table. The index type describes how SQL Server stores the index internally.

2. Why are indexes required in SQL Server?

Queries employ indexes to discover data from tables quickly. Tables and views both have indexes. The index on a table or view is quite similar to the index in a book.

If a book doesn't contain an index and we're asked to find a certain chapter, we'll have to browse through the whole book, beginning with the first page. If we have the index, on the other hand, we look up the chapter's page number in the index and then proceed to that page number to find the chapter.

Table and View indexes can help the query discover data fast in the same way. In reality, the presence of the appropriate indexes may significantly enhance query performance. If there is no index to aid the query, the query engine will go over each row in the table from beginning to end. This is referred to as a Table Scan, and the performance of a table scan is poor.

3. What are the different types of indexes in SQL Server?

- Clustered Index
- Non-Clustered Index

4. What is a Clustered Index?

In the case of a clustered index, the data in the index table will be arranged the same way as the data in the real table.

The index, for example, is where we discover the beginning of a book. The term "clustered table" refers to a table that has a clustered index.

The data rows in a table without a clustered index are kept unordered. A table can only have one clustered index, which is constructed when the table's main key constraint is invoked.

A clustered index determines the physical order of data in a table. As a result, a table can only have one clustered index.

5. What is a non-clustered index?

In a non-clustered index, the data in the index table will be organized differently than the data in the real database. A non-clustered index is similar to a textbook index. The data is kept in one location, while the index is kept in another. The index will contain references to the data's storage place.

A table can contain more than one non-clustered index since the non-clustered index is kept independently from the actual data, similar to how a book can have an index by chapters at the beginning and another index by common phrases at the conclusion.

The data is stored in the index in ascending or descending order of the index key, which has no bearing on data storage in the table. We can define a maximum of 249 non clustered indexes in a database.

6. In SQL Server, what is the difference between a clustered and a non-clustered index?

One of the most common SQL Server Indexes Interview Questions is this one. Let's look at the differences. There can only be one clustered index per table, although several non-clustered indexes can be.

The Clustered Index is quicker than the Non-Clustered Index by a little margin. When a Non-Clustered Index is used, an extra lookup from the Non-Clustered Index to the table is required to retrieve the actual data. A clustered index defines the row storage order in the database and does not require additional disk space. Still, a non-clustered index is kept independently from the table and thus requires additional storage space.

A clustered index is a sort of index that reorders the actual storage of entries in a table. As a result, a table can only have one clustered index. A non-clustered index is one in which the logical order of the index differs from the physical order in which the rows are written.

7. What is a SQL Server Unique Index?

If the "UNIQUE" option is used to build the index, the column on which the index is formed will not allow duplicate values, acting as a unique constraint. Unique clustered or unique non-clustered constraints are both possible.

If clustered or non-clustered is not provided when building an index, it will be non-clustered by default. A unique index is used to ensure that key values in the index are unique.

8. When does SQL Server make use of indexes?

SQL Server utilizes a table's indexes if the select, update, or delete statement included a "WHERE" condition and the where condition field was an indexed column. If an "ORDER BY" phrase is included in the select statement, indexes will be used as well.

Note: When SQL Server searches the database for information, it first determines the optimum execution plan for retrieving the data and then employs that plan, a full-page scan or an index scan.

9. When should a table's indexes be created?

If a table column is regularly used in a condition or order by clause, we must establish an index on it. It is not recommended that an index be created for each column since many indexes might reduce database performance. Any change to the data should be reflected in all index tables.

10. What is the maximum number of clustered and non-clustered indexes per table?

Clustered Index: Each table has only one Clustered Index. A clustered index stores all of the data for a single table, ordered by the index key. The Phone Book exemplifies the Clustered Index.

Non-Clustered Index: Each table can include many Non-Clustered Indexes. A Non-Clustered Index is an index found in the back of a book.

- 1 Clustered Index + 249 Nonclustered Index = 250 Index in SQL Server 2005
- 1 Clustered Index + 999 Nonclustered Index = 1000 Index in SQL Server 2008.

11. Clustered or non-clustered index, which is faster?

The Clustered Index is quicker than the Non-Clustered Index by a little margin. When a Non-Clustered Index is used, an extra lookup from the Non-Clustered Index to the table is required to retrieve the actual data.

12. In SQL Server, what is a Composite Index? What are the benefits of utilizing a SQL Server Composite Index? What exactly is a Covering Query?

A composite index is a two-or-more-column index, and composite indexes can be both clustered and non-clustered. A covering query is one in which all of the information can be acquired from an index. A clustered index always covers a query if chosen by the query optimizer because it contains all the data in a table.

13. What are the various index settings available for a table?

One of the following index configurations can be applied to a table:

- There are no indexes.

- A clustered index
- Many non-clustered indexes and a clustered index
- A non-clustered index
- Many non-clustered indexes

14. What is the table's name with neither a Cluster nor a Noncluster Index? What is the purpose of it?

Heap or unindexed table Heap is the name given to it by Microsoft Press Books and Book On-Line (BOL). A heap is a table that does not have a clustered index and does not have pointers connecting the pages. The only structures that connect the pages in a table are the IAM pages.

Unindexed tables are ideal for storing data quickly. It is often preferable to remove all indexes from a table before doing a large number of inserts and then to restore those indexes.

Data Integrity
1. What is data integrity?

The total correctness, completeness, and consistency of data are known as data integrity. Data integrity also refers to the data's safety and security in regulatory compliance, such as GDPR compliance. It is kept up-to-date by a set of processes, regulations, and standards that were put in place during the design phase. The information in a database will stay full, accurate, and dependable no matter how long it is held or how often it is accessed if the data integrity is protected.

The importance of data integrity in defending oneself against data loss or a data leak cannot be overstated: you must first guarantee that internal users are handling data appropriately to keep your data secure from harmful outside influences. You can ensure that sensitive data is never miscategorized or stored wrongly by implementing suitable data validation and error checking, therefore exposing you to possible danger.

2. What are data Integrity Types?

There must be a proper understanding of the two forms of data integrity, physical and logical, for maintaining data integrity. Both hierarchical and relational databases are collections of procedures and methods that maintain data integrity.

3. What is Physical Integrity?

Physical integrity refers to safeguarding data's completeness and correctness during storage and retrieval. Physical integrity is jeopardized when natural calamities hit, electricity goes out, or hackers interrupt database functionality. Data processing managers, system programmers, applications programmers, and internal auditors may be unable to access correct data due to human mistakes, storage degradation, and many other difficulties.

4. what is Logical Integrity?

In a relational database, logical integrity ensures that data remains intact when utilized in various ways. Logical integrity, like physical integrity, protects data from human mistakes and hackers, but differently. Logic integrity may be divided into four categories:

5. Explain the Integrity of entities

Entity integrity relies on generating primary keys to guarantee that data isn't shown more than once and that no field in a database is null. These unique values identify pieces of data. It's a characteristic of relational systems, which store data in tables that may be connected and used in many ways.

6. What is Referential Consistency?

The term "referential integrity" refers to a set of procedures that ensure that data is saved and utilized consistently. Only appropriate modifications, additions, or deletions of data are made, thanks to rules in the database's structure concerning how foreign keys are utilized. Rules may contain limits that prevent redundant data input, ensure proper data entry, and prohibit entering data that does not apply.

7. What is Domain Integrity?

Domain integrity is a set of operations that ensures that each piece of data in a domain is accurate. A domain is a set of permitted values that a column can hold in this context. Constraints and other measures that limit the format, kind, and amount of data submitted might be included.

8. User-defined integrity

User-defined integrity refers to the rules and limitations that users create to meet their requirements. When it comes to data security, entity, referential, and domain integrity aren't always adequate, and business rules must frequently be considered and included in data integrity safeguards.

9. What are the risks to data integrity?

The integrity of data recorded in a database can be affected for many reasons. The following are a few examples:

- Human error: Data integrity is jeopardized when people enter information erroneously, duplicate or delete data, fail to follow proper protocols, or make mistakes when implementing procedures designed to protect data.
- A transfer error occurs when data cannot be correctly transferred from one point in a database. In a relational database, transfer

errors occur when data is present in the destination table but not in the source table.

- Viruses and bugs: Spyware, malware, and viruses are programs that can infiltrate a computer and change, erase, or steal data.
- Sudden computer or server breakdowns, as well as issues with how a computer or other device performs, are instances of serious failures that might indicate that your hardware has been hacked. Compromise hardware might cause data to be rendered inaccurately or incompletely. Also, they might limit or reduce data access or make information difficult to utilize.

The following steps can be taken to reduce or remove data integrity risks:

- Limiting data access and modifying permissions to prevent unauthorized parties from making changes to data
- Validating data, both when it's collected and utilized, ensures that it's accurate.
- Using logs to track when data is added, edited, or removed is a good way to back up data.
- Internal audits are carried out regularly.
- Using software to spot errors

SQL Cursor
1. What is a cursor in SQL Server?

A cursor is a database object that represents a result set and handles data one row at a time.

2. How to utilize the Transact-SQL Cursor

Make a cursor declaration, Activate the cursor, Row by row, get the data. Deallocate cursor, Close cursor.

3. Define the different sorts of cursor locks

There are three different types of locks. ONLY READ: This stops the table from being updated.

4. Tips for cursor optimization

When not in use, close the pointer. Remember to deallocate the cursor after closing it.

5. The cursor's disadvantages and limitations

Cursor consumes network resources by requiring a round-trip each time it pulls a record.

7

DATA WRANGLING

Data wrangling, also known as data munging, is the act of changing and mapping data from one "raw" data type into another to make it more suitable and profitable for downstream applications such as analytics. Data wrangling aims to ensure that the data is high quality and usable. Data analysts generally spend the bulk of their time wrangling data rather than analyzing it when it comes to data analysis.

Further munging, data visualization, data aggregation, training a statistical model, and many more possible uses are all part of the data wrangling process. Data wrangling often involves extracting raw data from a data source, "munging" or parsing the raw data into preset data structures, and ultimately depositing the generated information into a data sink for storage and future use.

Benefits

With rawer data emanating from more data that isn't intrinsically usable, more effort is needed in cleaning and organizing before it can be evaluated. This is where data wrangling comes into play. The output of data wrangling can give useful metadata statistics for gaining more insights into the data; nevertheless, it is critical to ensure that information is consistent since inconsistent metadata might present bottlenecks. Data wrangling enables analysts to swiftly examine more complex data, provide more accurate results, and make better judgments as a result. Because of its results, many firms have shifted to data wrangling systems.

The Basic Concepts

The following are the major steps in data wrangling:

- **Discovering**

The first step in data wrangling is to obtain a deeper knowledge: different data types are processed and structured differently.

- **Structuring**

This step involves arranging the information. Raw data is frequently disorganized, and most of it may be useless in the final output. This step is necessary for the subsequent calculation and analysis to be as simple as possible.

- **Cleaning**

Cleaning data can take various forms, such as identifying dates that have been formatted incorrectly, deleting outliers that distort findings, and formatting null values. This phase is critical for ensuring the data's overall quality.

- **Enriching**

Determine whether more data would enrich the data set and could be easily contributed at this point.

- **Validating**

This process is comparable to cleaning and structuring. To ensure data consistency, quality, and security, use recurring sequences of validation rules. An example of a validation rule is confirming the correctness of fields through cross-checking data.

- **Publishing**

Prepare the data set for usage in the future. This might be done through software or by an individual. During the wrangling process, make a note of any steps and logic.

This iterative procedure should result in a clean and useful data collection that can be analyzed. This is a time-consuming yet beneficial technique since it helps analysts extract information from a large quantity of data that would otherwise be difficult.

Typical Use

Extractions, parsing, joining, standardizing, augmenting, cleansing, consolidating, and filtering is common data transformations applied to distinct entities (e.g., fields, rows, columns, data values, etc.) within a data set. They can include actions like extractions, parsing, joining, standardizing, augmenting, cleansing, consolidating, and filtering to produce desired wrangling outputs that can be leveraged.

Individuals who will study the data further, business users who will consume the data directly in reports, or systems that will further analyze it and write it to targets such as data warehouses, data lakes, or downstream applications might be the receivers.

Data cleaning is the process of detecting and correcting (or removing) corrupt or inaccurate data from a recordset, table, or database, and it entails identifying incomplete, incorrect, inaccurate, or irrelevant parts of the data and then replacing, modifying, or deleting the dirty or coarse data. Data purification may be done in real-time using data wrangling tools or in batches using scripting.

After cleaning, a data set should be consistent with other similar data sets in the system. User entry mistakes, transmission or storage damage, or differing data dictionary definitions of comparable entities in various stores may have caused the discrepancies discovered or eliminated. Data cleaning varies from data validation in that validation nearly always results in data being rejected from the system at the time of entry. In contrast, data cleaning is conducted on individual records rather than batches.

Data cleaning may entail repairing typographical mistakes or verifying and correcting information against a known set of entities. Validation can be stringent (e.g., rejecting any address without a valid postal code) or fuzzy (e.g., approximating string matches) (such as correcting records that partially match existing, known records). Some data cleaning software cleans data by comparing it to an approved data collection. Data augmentation is a popular data cleaning process, and data is made more comprehensive by adding relevant information— appending addresses with phone numbers associated with that address. Data cleaning can also include data harmonization (or normalization). Data harmonization is the act of combining data from "different file formats, naming conventions, and columns" and transforming it into a single, coherent data collection; an example is the extension of abbreviations ("st, rd, etc." to "street, road, etcetera").

Data Quality

A set of quality requirements must be met for data to be considered high-quality. These are some of them:

- Validity (statistics): The degree to which the measurements adhere to established business norms or limits. Validity is rather straightforward to verify when contemporary database technology is utilized to create data-capture systems. Invalid data is most commonly found in legacy situations or when the wrong data-capture technique was employed. The following are the types of data constraints:

- Data-Type Constraints — For example, values in a column must be of a specific data type, such as Boolean, numeric (integer or real), date, and so on.
- Range Constraints: Numbers or dates should normally fall inside a specific range. That is, they have permitted minimum and maximum values.
- Mandatory Constraints: Some columns can't be left blank.
- Unique Constraints: Within a dataset, a field, or a combination of fields, must be unique. No two people can have the same social security number, for example.
- Set-Membership constraints: A column's values are derived from a collection of discrete values or codes. A person's sex, for example, might be Female, Male, or Non-Binary.
- Foreign-key constraints: This is a more general instance of set membership. The set of values is defined in a unique values column in another table. For example, in a US taxpayer database, the "state" field must be one of the US's specified states or territories; the list of permitted states/territories is kept in a separate State table. The phrase "foreign key" comes from the world of relational databases.
- Regular expression patterns: Text fields will occasionally need to be verified using regular expression patterns. Phone numbers, for example, may be required to follow the pattern (999) 999-9999.
- Cross-field validation: Certain constraints must be met if several fields are used. For example, in laboratory medicine, the total of

the differential white blood cell count components must equal 100. (since they are all percentages). A patient's discharge date from the hospital cannot be earlier than the date of admission in a hospital database.

- Accuracy: This measures the degree of conformance to a standard or real value. In the general instance, accuracy is difficult to accomplish through data cleaning. It necessitates access to an external source of data that includes the correct value: such "gold standard" data is frequently unavailable. External databases that match up zip codes to geographical locations (city and state) and also assist verify that street addresses inside these zip codes exist have been used to achieve accuracy in specific cleaning scenarios, especially customer contact data.

- Completeness: Refers to how well all of the required measurements are known. The data cleaning technique can't correct inconsistencies since it can't deduce facts that weren't documented when the data in question was first recorded. Suppose a system requires that some columns not be empty. In that case, one may work around the difficulty by specifying a value that signals "unknown" or "missing," but providing default values does not guarantee that the data is full.)

- Inconsistency: The degree to which a collection of measurements is equal across systems. When two data elements in a data collection contradict one other, inconsistency develops. For example, only one may be correct if a client is stored in two distinct systems with two different current addresses. Fixing inconsistency is not always possible: it necessitates a variety of strategies, such as determining which data were recorded more recently, determining which data source is likely to be the most reliable (the latter knowledge may be unique to a given organization), or simply trying to find the truth by testing both data items (e.g., calling up the customer).

- Uniformity: The degree to which a collection of data measures is stated in all systems using the same units of measure. Weight may be reported in pounds or kilograms in datasets compiled from many locations and must be transformed to a single measure via an arithmetic transformation.

- Integrity: Because it is inadequately explicit, the term integrity is seldom employed by itself in data-cleaning situations. It comprises correctness, consistency, and some validation elements (see also data integrity). (For example, the enforcement of foreign-key limitations is referred to as "referential integrity.").

Process

1. Data auditing: Anomalies and inconsistencies are detected using statistical and database approaches, which leads to identifying the anomalies' features and locations. Several commercial software programs allow you to specify various types of constraints (using a syntax similar to that of a normal programming language, such as JavaScript or Visual Basic) and then produce code that examines the data for violations of these constraints. This process is defined by "workflow definition" and "workflow execution." Microcomputer database packages such as Microsoft Access or File Maker Pro will let you perform such checks interactively on a constraint-by-constraint basis.

2. Workflow specification: The identification and elimination of anomalies are carried out by a workflow, which is a series of activities on data. It is specified after the data auditing process and is critical in creating a high-quality data result. The reasons for abnormalities and inaccuracies in the data must be carefully evaluated to establish a suitable workflow.

3. Workflow execution: After the workflow's definition is

complete and its validity is checked, the workflow is run at this step. The process should be efficient, even when working with massive volumes of data, which is inherently a trade-off because performing a data-cleaning procedure might be computationally expensive.

4. Post-processing and controlling: After the cleaning operation has been completed, the results are adequately examined. If possible, data that could not be updated during the workflow execution is manually repaired. Consequently, the data is verified again in the data-cleaning process, allowing the definition of an extra workflow to purify the data through automated processing further.

Good quality source data is linked to a company's "Data Quality Culture," which must start at the top. It's not merely a question of putting in place robust validation checks on input screens because users can often get around these tests, no matter how powerful they are. There is a nine-step process for firms looking to increase data quality:

- Make a strong commitment to a data quality culture.
- Push for process reengineering at the executive level.
- Invest in improving the data entry environment.
- Invest in improving application integration.
- Spend money to alter the way processes are carried out.
- Encourage all team members to be aware of the situation from beginning to end.
- Encourage cross-departmental collaboration.
- Extol the virtues of data quality in public.
- Measure and enhance data quality regularly.

Other options include:

- Parsing: Parsing is used to discover syntax problems. A parser determines if a string of data falls inside the parameters of the data specification. A parser similarly deals with grammar and languages.

- Data transformation: Data transformation is converting data from one format to another so that the appropriate application may read it. Value conversions or translation functions and normalizing numeric numbers to adhere to minimum and maximum values fall under this category.

- Duplicate elimination: Duplicate detection necessitates a technique that determines if data includes duplicate representations of the same object. Data is usually sorted using a key that brings duplicate entries closer to easier detection.

- Statistical methods: By evaluating the data with mean, standard deviation, range, or clustering techniques, an expert might discover numbers that are unexpected and hence incorrect. Although it is difficult to fix such data since the real value is unknown, the problem can be remedied by adjusting the numbers to an average or other statistical value. Missing values that can be replaced by one or more reasonable values, normally produced using complex data augmentation processes, can also be handled using statistical approaches.

The graphic depiction of data is the subject of data visualization, which is an interdisciplinary discipline. When the data is large, such as in a time series, it is a very effective communication.

Data Visualization
1. What constitutes good data visualization?

- Use of color theory
- Data positioning
- Bars over circles and squares
- Reducing chart junk by avoiding 3D charts and eliminating the use of pie charts to show proportions

2. How can you see more than three dimensions in a single chart?

Typically, data is shown in charts using height, width, and depth in pictures; however, to visualize more than three dimensions, we employ visual cues such as color, size, form, and animations to portray changes over time.

3. What processes are involved in the 3D Transformation of data visualization?

Data transformation in 3D is necessary because it provides a more comprehensive picture of the data and the ability to see it in more detail.

The overall procedure is as follows:

- Viewing Transformation
- Workstation Transformation
- Modeling Transformation
- Projection Transformation

4. What is the definition of Row-Level Security?

Row-level security limits the data a person can see and access based on their access filters. Depending on the visualization tool being used, users can specify row-level security. Several prominent visualization technologies, including Qlik, Tableau, and Power BI, are available.

5. What Is Visualization "Depth Cueing"?

Depth cueing is a fundamental challenge in vision approaches. Some 3D objects lack visual line and surface identification due to a lack of depth information. To draw attention to the visible lines, draw them as dashed lines and delete the unseen ones.

6. Explain Surface Rendering in Visualization?

- Lightening conditions in the screen

- Degree of transparency
- Assigned characteristics
- Exploded and cutaway views
- How rough or smooth the surfaces are to be
- Three dimensional and stereoscopic views

7. What is Informational Visualization?

Information visualization focused on computer-assisted tools to study huge amounts of abstract data. The User Interface Research Group at Xerox PARC, which includes Dr. Jock Mackinlay, was the first to develop the phrase "information visualization." Selecting, manipulating, and showing abstract data in a way that allows human engagement for exploration and comprehension is a practical use of information visualization in computer applications. The dynamics of visual representation and interaction are important features of information visualization. Strong approaches allow the user to make real-time changes to the display, allowing for unequaled observation of patterns and structural relationships in abstract data.

8. What are the benefits of using Electrostatic Plotters?

- They outperform pen plotters and high-end printers in terms of speed and quality.
- A scan-conversion feature is now available on several electrostatic plotters.
- There are color electrostatic plotters on the market, and they make numerous passes over the page to plot color images.

9. What is Pixel Phasing?

Pixel phasing is an antialiasing method that smooths out stair steps by shifting the electron beam closer to the places defined by the object shape.

10. Define Perspective Projection

This is accomplished by projecting points to the display plane and converging points. As a result, items further away from the viewing point should be smaller than those present here.

11. Explain winding numbers in visualization

The winding number approach determines whether a particular point is inside or outside the polygon. This approach gives all the edges that cross the scan line a direction number. If the edge begins below the line and finishes above the scan line, the direction should be -1; otherwise, it should be 1. When the value of the winding number is nonzero, the point is considered to be inside polygons or two-dimensional objects.

12. What is Parallel Projection?

Parallel projection is the process of creating a 2D representation of a 3D scene—project points from the object's surface along parallel lines on the display plane. Different 2D perspectives of things may be created by projecting the visible spots.

13. What is a blobby object?

Some objects may not retain a constant form but instead vary their surface features in response to particular motions or close contact with other objects. Molecular structures and water droplets are two examples of blobby objects.

14. What is Non-Emissive?

They are optical effects that turn light from any source into pictorial forms, such as sunshine. A good example is the liquid crystal display.

15. What is Emissive?

Electrical energy is converted into light energy by the emissive display.

Examples include plasma screens and thin film electroluminescent displays.

16. What is Scan Code?

When a key is pushed on the keyboard, the keyboard controller stores a code corresponding to the pressed key in the keyboard buffer, which is a section of memory. The scan code is the name given to this code.

17. What is the difference between a window port and a viewport?

A window port refers to a section of an image that a window will display. The viewport is the display area of the selected portion of the form in which the selected component is displayed.

8

DATA SCIENCE INTERVIEW EXTRA

Depending on the firm and sector, data science interview practices might differ. They usually begin with a phone interview with the recruiting manager, followed by more onsite interviews. You'll be asked technical and behavioral data science interview questions, and you'll almost certainly have to complete a skills-related project. Before each interview, you should examine your CV and portfolio and prepare for possible interview questions.

Data science interview questions will put your knowledge and abilities in statistics, programming, mathematics, and data modelling to the test. Employers will consider your technical and soft talents, as well as how well you might fit into their organization.

If you prepare some typical data science interview questions and responses, you can confidently go into the interview. During your data science interview, you may anticipate being asked different kinds of data scientist questions.

Extra Interview Questions

Employers are searching for well-versed applicants in data science ideas and practices. Depending on the profession and abilities necessary, data-related interview questions will differ.

Here are a few extra interview questions and responses:

1. What is the distinction between deep learning and machine learning?

2. Give a detailed explanation of the Decision Tree algorithm.

3. What exactly is sampling? How many different sampling techniques are you familiar with?

4. What is the distinction between a type I and a type II error?

5. What is the definition of linear regression? What are the definitions of the words p-value, coefficient, and r-squared value? What are the functions of each of these elements?

6. What is statistical interaction?

7. What is selection bias?

8. What does a data set with a non-Gaussian distribution look like?

9. What is the Binomial Probability Formula, and how does it work?

10. What distinguishes k-NN clustering from k-means clustering?

11. What steps would you take to build a logistic regression model?

12. Explain the 80/20 rule and its significance in model validation.

13. Explain the concepts of accuracy and recall. What is their relationship to the ROC curve?

14. Distinguish between the L1 and L2 regularization approaches.

15. What is root cause analysis, and how does it work?

16. What is hash table collisions?

17. Before implementing machine learning algorithms, what are some procedures for data wrangling and cleaning?

18. What is the difference between a histogram and a box plot?

19. What is cross-validation, and how does it work?

20. Define the terms "false-positive" and "false-negative." Is it preferable to have a large number of false positives or a large number of false negatives?

21. In your opinion, which is essential, model performance or accuracy, when constructing a machine learning model?

22. What are some examples of scenarios in which a general linear model fails?

23. Do you believe that 50 little decision trees are preferable to a single huge one? Why?

Interview Questions on Technical Abilities

In a data science interview, technical skills questions are used to assess your data science knowledge, skills, and abilities. These questions will be connected to the Data Scientist's unique work duties.

You should demonstrate your thought process and clearly explain how you arrived at a solution when addressing issues.

The following are some examples of technical data science interview questions:

1. What are the most important data scientist tools and technical skills?

Because data science is such a sophisticated profession, you'll want to demonstrate to the hiring manager that you're familiar with all of the most up-to-date industry-standard tools, software, and programming languages. Data scientists typically use R and Python among the different statistical programming languages used in data research. Both may be used for statistical tasks, including building a nonlinear or linear model, regression analysis, statistical testing, data mining, and so on. RStudio Server is another essential data science application, whereas Jupyter Notebook is frequently used for statistical modelling, data visualizations, and machine learning functions, among other things. Tableau, PowerBI, Bokeh, Plotly, and Infogram are just a few of the dedicated data visualization tools that Data Scientists use frequently. Data scientists must also have a lot of SQL and Excel skills.

"Any specific equipment or technical skills required for the position you're interviewing for should also be included in your response. Examine the job description, and if there are any tools or applications you haven't used before, it's a good idea to familiarize yourself with them before the interview."

2. How should outlier values be treated?

Outliers can be eliminated in some cases. You can remove garbage values or values that you know aren't true. Outliers with extreme values that differ significantly from the rest of the data points in a collection can also be deleted. Suppose you can't get rid of outliers. In that case, you may rethink whether you choose the proper model, employ methods (such random forests) that aren't as affected by outlier values, or attempt normalizing your data.

Extras

3. Tell me about a unique algorithm you came up with.

4. What are some of the advantages and disadvantages of your preferred statistics software?

5. Describe a data science project where you had to work with a lot of code. What did you take away from the experience?

6. How would you use five dimensions to portray data properly?

7. Assume using multiple regression to create a predictive model. Describe how you plan to test this model.

9. How do you know that your modifications are better than doing nothing while updating an algorithm?

10. What would you do if you had an unbalanced data set for prediction (i.e., many more negative classes than positive classes)?

11. How would you validate a model you constructed using multiple regression to produce a predictive model of a quantitative outcome variable?

12. I have two equivalent models of accuracy and processing power. Which one should I use for production, and why should I do so?

13. You've been handed a data set with variables that have more than 30% missing values. What are your plans for dealing with them?

Interview on Personal Concerns

Employers will likely ask generic questions to get to know you better, in addition to assessing your data science knowledge and abilities. These questions will allow them to learn more about your work ethic, personality, and how you could fit into their business culture.

Here are some personal Data Scientist questions that can be asked:

1. What qualities do you believe a competent Data Scientist should possess?

Your response to this question will reveal a lot about how you view your position and the value you offer to a company to a hiring manager. In your response, you might discuss how data science necessitates a unique set of competencies and skills. A skilled Data Scientist must be able to combine technical skills like parsing data and creating models with business sense like understanding the challenges they're dealing with and recognizing actionable insights in their data. You might also include a Data Scientist you like in your response, whether it's a colleague you know or an influential industry figure.

Extras

2. Tell me more about yourself.

3. What are some of your strengths and weaknesses?

4. Which data scientist do you aspire to be the most like?

5. What attracted you to data science in the first place?

6. What unique skills do you believe you can provide to the team?

7. What made you leave your last job?

8. What sort of compensation/pay do you expect?

9. Give a few instances of data science best practices.

10. What data science project at our organization would you want to work on?

11. Do you like to work alone or in a group of Data Scientists?

12. In five years, where do you see yourself?

13. How do you deal with tense situations?

14. What inspires and motivates you?

15. What criteria do you use to determine success?

16. What kind of work atmosphere do you want to be in?

17. What do you enjoy doing outside of data science?

Interview Questions on Communication and Leadership

Data scientists need to be able to lead and communicate effectively. Employers prize applicants who can take the initiative, share their knowledge with colleagues, and articulate data science goals and plans.

Here are some examples of data science interview questions for leadership and communication:

1. Tell me about a time when you were a multi-disciplinary team member.

A Data Scientist works with a diverse group of people in technical and non-technical capacities. Working with developers, designers, product experts, data analysts, sales and marketing teams, and top-level executives, not to mention clients, is not unusual for a Data Scientist. So, in your response to this question, show that you're a team player who enjoys the opportunity to meet and interact with people from other departments. Choose a scenario in which you reported to the company's highest-ranking officials to demonstrate not just that you can communicate with anybody but also how important your data-driven insights have been in the past.

Extras

2. Could you tell me about a moment when you used your leadership skills on the job?

3. What steps do you use to resolve a conflict?

4. What method do you like to use to establish rapport with others?

5. Discuss a successful presentation you delivered and why you believe it went well.

6. How would you communicate a complex technical issue to a colleague or client who is less technical?

7. Describe a situation in which you had to be cautious when discussing sensitive information. How did you pull it off?

8. On a scale of 1 to 10, how good are your communication skills? Give instances of situations that prove the rating is correct.

Behavioral Interview Questions

Employers use behavioral interview questions to seek specific circumstances that demonstrate distinct talents. The interviewer wants to know how you handled previous circumstances, what you learned, and what you can add to their organization.

In a data science interview, behavioral questions could include:

1. Tell me about a moment when you were tasked with cleaning and organizing a large data collection.

According to studies, Data Scientists spend most of their time on data preparation rather than data mining or modelling. As a result, if you've worked as a Data Scientist before, you've almost certainly cleaned and organized a large data collection. It's also true that this is a job that just a few individuals like. However, data cleaning is one of the most crucial processes. As a result, you should walk the hiring manager through your data preparation process, including deleting duplicate observations, correcting structural problems, filtering outliers, dealing with missing data, and validating data.

Extras

2. Tell me about a data project you worked on and met a difficulty. What was your reaction?

3. Have you gone above and beyond your normal responsibilities? If so, how would you go about doing it?

4. Tell me about a period when you were unsuccessful and what you learned from it.

5. How have you used data to improve a customer's or stakeholder's experience?

6. Give me an example of a goal you've attained and how you got there.

7. Give an example of a goal you didn't achieve and how you dealt with it.

8. What strategies did you use to meet a tight deadline?

9. Tell me about an instance when you successfully settled a disagreement.

Interview Questions Top Companies

I collected data science interview questions from some of the biggest IT firms to give you an idea of what you could be asked in an interview.

1. What's the difference between support vector machines and logistic regression? What is an example of when you would choose to use one

over the other?

2. What is the integral representation of a ROC area under the curve?

3. A disc is spinning on a spindle, and you don't know which direction the disc is spinning. A set of pins is given to you. How will you utilize the pins to show how the disc is spinning?

3. What would you do if you discovered that eliminating missing values from a dataset resulted in bias?

4. What metrics would you consider when addressing queries about a product's health, growth, or engagement?

5. When attempting to address business difficulties with our product, what metrics would you consider?

6. How would you know whether a product is performing well or not?

7. What is the best way to tell if a new observation is an outlier? What is the difference between a bias-variance trade-off?

8. Discuss how to randomly choose a sample of a product's users.

9. Before using machine learning algorithms, explain the data wrangling and cleaning methods.

10. How would you deal with a binary classification that isn't balanced?

11. What makes a good data visualization different from a bad one?

12. What's the best way to find percentiles? Write the code for it.

13. Make a function that determines whether a word is a palindrome.

CONCLUSION

Being a data scientist is steadily becoming a popular job; many data scientist jobs are now available worldwide and grow significantly every year. According to Harvard Business Review, it's the *"sexiest job of the twenty-first century"*, and the expanding employment trend in the field seems to back up that claim.

The data scientist interview process may be quite wide and complicated. Because your job might cover a wide range of topics (depending on the organization you work for), the questions you will be asked during an interview will be rather varied. For instance, you may be asked questions on statistics, SQL, and machine learning in an interview, as well as questions about coding, algebra, and programming.

In this interview guide, I examined a database of genuine interview questions from real firms that we have amassed over time. These questions were used to examine what a corporate interview entails, and I have gone through all of the pertinent questions and provided solutions.

Like other professional interviews, data science interviews need a lot of planning. To guarantee that you are prepared for back-to-back questions on statistics, programming, and machine learning, you must master a variety of disciplines.

Firms perform different sorts of data science interviews. Some data science interviews are heavily focused on products and metrics. These interviews focus primarily on product topics; such as what metrics you would use to illustrate where a product may be improved. These are frequently used in conjunction with SQL and Python questions. Another form of data science interview combines programming and machine learning.

If you are unsure about the sort of interview you will have, I recommend asking the recruiter. Some organizations excel at maintaining consistency in interviews, but even then, teams might vary based on what they are looking for. Seek opinions, ask questions, and cover as much as possible to secure that position. Good Luck!!